Communications in Computer and Information Science 480

More information about this series at http://www.springer.com/series/7899

Dongyan Zhao · Jianfeng Du
Haofen Wang · Peng Wang
Donghong Ji · Jeff Z. Pan (Eds.)

The Semantic Web and Web Science

8th Chinese Conference, CSWS 2014
Wuhan, China, August 8–12, 2014
Revised Selected Papers

 Springer

Editors

Dongyan Zhao
Peking University
Beijing
China

Peng Wang
Southeast University
Nanjing
China

Jianfeng Du
Guangdong University of Foreign Studies
Guangzhou
China

Donghong Ji
Wuhan University
Wuhan
China

Haofen Wang
East China University
Shanghai
China

Jeff Z. Pan
The University of Aberdeen
Aberdeen
UK

ISSN 1865-0929   ISSN 1865-0937   (electronic)
Communications in Computer and Information Science
ISBN 978-3-662-45494-7   ISBN 978-3-662-45495-4   (eBook)
DOI 10.1007/978-3-662-45495-4

Library of Congress Control Number: 2014955539

Springer Heidelberg New York Dordrecht London

Printed on acid-free paper

Springer-Verlag GmbH Berlin Heidelberg is part of Springer Science+Business Media
(www.springer.com)

Preface

With only a few years, the rapidly developingWeb has led to revolutionary changes in the whole of human society. The SemanticWeb is the next generation of Web. It provides a common framework that allows data to be shared and reused across applications, enterprises, and community boundaries. Semantic Web technology facilitates building a large-scale Web of machine-readable knowledge that allows for data reuse and integration. Nowadays, many scholars and experts are devoting themselves to the work of applying the SemanticWeb theories into specific practice, while in turn improving the Semantic Web theories and techniques according to the demand in practice. Web Science involves the full scope ofWeb-related applications, integrating all directions of Web-related interdisciplinary research into a new field. The jointly held Chinese Semantic Web Symposium and Chinese Web Science Conference aim to promote expansions from the Semantic Web to Web Science, and to discuss the core technology on the Next Generation Web, including swarm intelligence, semantic search, Web security, natural language processing (NLP), etc.

Following the success of the Seventh Chinese Semantic Web Symposium and the Second Chinese Web Science Conference in Shanghai, China, we organized the Eighth Chinese Semantic Web and Web Science Conference (CSWS 2014) in Wuhan, China. The theme of CSWS 2014 was "Semantics and Big Data." A summer school of three days was organized before the main conference. On the first day, an expert in semantic search, Dr. Haofen Wang, from East China University of Science and Technology, gave lectures on semantic search on linked data. In the morning session of the second day, an expert from Tonglian Data Company, Dr. Long Jiang, gave lectures on algorithms for matching and ranking Internet advertisements. In the afternoon section of the second day, an expert from Baidu Company, Dr. Yanjun Ma, gave a talk on deep question and answering and its application to search. In the morning section of the last day, an expert in Big Data, Prof. Ping Li, from Rutgers University, USA, gave a tutorial titled "BigData: Hashing Algorithms for Large-Scale Search, Learning, and Compressed Sensing." In the afternoon section of the last day, Dr. Guohui Xiao from Free University of Bozen-Bolzano, Italy and Dr. Dafu Deng from Tengxun Company, gave talks on ontology-based data access and on big data-based game marketing reformation, respectively.

This volume contains the papers presented at CSWS 2014. The conference received 61 submissions. Each submission was assigned to at least three Program Committee (PC) members to review. After a rigorous reviewing process, 22 research papers were selected for publication in the proceedings, and seven research papers were selected for publication in the Journal of Southeast University (Chinese Edition) and Journal of Wuhan University (Natural Science Chinese Edition). These submissions cover a wide range of topics, including Ontology Reasoning and Learning, Semantic Data Generation and Management, Semantic Technology Applications, etc.

We would like to thank the excellent work of the PC. Each PC member was assigned three to four papers to review and their timely and professional reviews were helpful for us to select submissions of high quality. We also thank everyone who was involved in organizing the conference, especially Prof. Guilin Qi, Prof. Jinguang Gu, Prof. Siwei Yu, Prof. Donghong Ji, Prof. Jeff Z. Pan, Prof. Huajun Chen, Prof. Zhiqiang Gao, Prof. Jie Tang, Prof. Wei Hu, and Prof. Zhichun Wang.

September 2014

Dongyan Zhao
Jianfeng Du
Haofen Wang
Peng Wang
Donghong Ji
Jeff Z. Pan

Organization

Organization Committee

Conference Chairs

Donghong Ji   Wuhan University, China
Jeff Z. Pan   University of Aberdeen, UK

Program Committee Chairs

Dongyan Zhao   Peking University, China
Jianfeng Du   Guangdong University of Foreign Studies, China

Workshop Chair

Zhichun Wang   Beijing Normal University, China

Publicity Chair

Wei Hu   Nanjing University, China

Publication Chair

Zhiqiang Gao   Southeast University, China

Poster and Demo Chairs

Haofeng Wang   East China University of Science and Technology, . China
Peng Wang   Southeast University, China

Summer School Chairs

Guilin Qi   Southeast University, China
Jie Tang   Tsinghua University, China

Tutorial Chair

Huajun Chen   Zhejiang University, China

Local Chair

Jinguang Gu   Wuhan University of Science and Technology,
                               China
Siwei Yu   Wuhan University, China

Program Committee

Dongyan Zhao   Peking University, China
Jianfeng Du   Guangdong University of Foreign Studies, China
Yue Ma   Theoretical Computer Science and TU Dresden,
                               Germany
Zhixing Li   Chongqing University, Germany
Xin Xin   Beijing Institute of Technology, China
Jinguang Gu   Wuhan University of Science and Technology,
                               China
Peng Wang   Southeast University, China
Yueguo Chen   Renmin University, China
Yuzhong Qu   Nanjing University, China
Guilin Qi   Southeast University, China
Huajun Chen   Zhejiang University, China
Haofen Wang   East China University of Science and Technology,
                               China
Kewen Wang   Griffth University, Australia
Zhiyuan Liu   Tsinghua University, China
Zhichun Wang   Beijing Normal University, China
Ming Zhang   Peking University, China
Guojie Song   Peking University, China
Yutao Ma   Wuhan University, China
Pingpeng Yuan   Huazhong University of Science and Technology,
                               China
Jie Tang   Tsinghua University, China
Zhiqiang Gao   Southeast University, China
Ruifeng Xu   Harbin Institute of Technology, China
Jiefeng Cheng   The Chinese University of Hong Kong,
                               Hong Kong
Kang Liu   Institute of Automation and Chinese Academy
                               of Sciences, China
Yuxiao Dong   University of Notre Dame, USA
Yingjie Li   Lehigh University, USA
Xianpei Han   Institute of Software and Chinese Academy
                               of Sciences, China
Tong Ruan   East China University of Science and Technology,
                               China
Bangyong Liang   Microsoft, China

Fangtao Li	Google Research, USA
Huaiyu Wan	Beijing Jiaotong University, China
Yuqing Zhai	Southeast University, China
Xin Wang	Tianjin University, China
Gang Wu	Northeastern University, USA
Dandan Song	Tsinghua University, China
Zhihong Chong	Southeast University, China
Ruixuan Li	Huazhong University of Science and Technology, China
Shenghui Wang	OCLC Research, The Netherlands
Liang Zhang	Fudan University, China
Huawei Shen	Institute of Computing Technology and Chinese Academy of Sciences, China
Yi-Dong Shen	Institute of Software and Chinese Academy of Sciences, China
Shu Zhao	University of California, USA
Patrick Lambrix	Linköping University, Sweden
Xiaoqin Xie	Harbin Engineering University, China
Zhisheng Huang	Vrije University of Amsterdam, The Netherlands
Xuhui Li	Wuhan University, China

Contents

Semantic Technology Applications

Ontology Reasoning and Learning

SPARQL Query Recommendation
for Exploring RDF Repositories

Boliang Chen[1]([✉]), Jing Mei[2], Wen Sun[2], Ruilong Su[1], Haofen Wang[3],
Gang Hu[2], Guotong Xie[2], and Yong Yu[1]

[1] Shanghai Jiao Tong University, Shanghai, China
blchen@apex.sjtu.edu.cn
[2] IBM China Research Lab, Beijing, China
[3] East China University of Science and Technology, Shanghai, China

Abstract. With the rapid development of Semantic Web, more and
more RDF repositories, such as Linking Open Data (LOD), are available on the web. Generally, there are two services provided for exploring
those RDF repositories, one is the keyword lookup, and the other is
the SPARQL endpoint. Most users choose the lookup service, and millions of web logs have been recorded. Although, users expect to submit more expressive queries than keyword lookup, the complexity of
SPARQL undoubtedly scared users away. This paper proposes a method
of SPARQL query recommendation for exploring RDF repositories. By
analyzing web logs of the lookup service, our method extracts the user
access patterns, which will be used to recommend SPARQL queries. We
implement our method based on Zhishi.me, a Chinese RDF repository
with about 150 million triples as well as over one-year web logs. We
believe the proposed method will further facilitate the SPARQL query
research.

Keywords: RDF repository · SPARQL query · Recommendation
system

1 Introduction

As one of the W3C specifications, RDF describes a standard data model for data
interchange on the Web. Through unremitting efforts in the past decades, more
and more data in form of RDF are published and interlinked with other semantic
data sources, which forms a open community called Linking Open Data (LOD).
Lots of users access these data via linked data queries. Running for years, many
existing RDF repositories collect a mass of open data in different domains as
well as user access patterns for various datasets. As a standard query language,
SPARQL has been wildly used for RDF repositories exploration.

In most cases, two kinds of services are provided for exploring those RDF
repositories, namely the keyword lookup and the SPARQL endpoint. In reality,
a majority of users prefer the lookup service, and their access behaviors have
been recorded in millions of web logs. Although users may expect to submit more

© Springer-Verlag Berlin Heidelberg 2014
D. Zhao et al. (Eds.): CSWS 2014, CCIS 480, pp. 3–16, 2014.
DOI: 10.1007/978-3-662-45495-4_1

expressive queries than simply keyword lookup in some cases, the complexity of SPARQL undoubtedly scared users away. Also we observe realize that for a new user wishing to explore RDF graphs, he may be more interested in traversing among multiple URIs/resources rather than looking into details of a randomly selected URI/resource in most cases. For example, when a new user sends a SPARQL query (e.g., select ?s ?p ?o where{ ?s ?p ?o }, also known as a simple "SPO" query) to explore an unfamiliar RDF repository, he may not get a proper size of result set (either too large or too small) at the first time. In the former case, he will be exposed to the flood of triples, resulting in finding it difficult to get "useful" information (as Fig. 3 shows). In the latter case, he may feel disappointed and has to send another SPARQL query to get more triples (as shown in Fig. 5).

To help users to find things that are interesting and narrow down the set of choices, recommendation systems have been well researched in academic and widely employed in industries [17]. Some basic techniques in recommendation systems, including collaborative filtering [7], content-based [22] and knowledge-based approaches [9], have been implemented and applied in various scenarios. Hence, we would like to utilize those methods to make a good recommendation for new users so that they can avoid receiving improper results and explore RDF repositories more comfortably using SPARQL endpoints.

In this paper, we propose a method of SPARQL query recommendation for exploring RDF repositories. We extract user access patterns through analyzing web logs of the lookup service so as to make SPARQL query recommendation. Our method is implemented on Zhishi.me [21], a Chinese RDF repository with about 150 million triples and over 1-year web logs. We have faith that the proposed method will further facilitate the SPARQL query research.

The organization of this paper is as follows. Section 2 describes the proposed framework. Sections 3 and 4 present technical details of the most important parts of our method. Our implementation on the Zhishi.me RDF repository and recommendation results are given in Sect. 5. Section 6 reviews the related work. Finally, Sect. 7 concludes this paper.

2   Framework

Figure 1 describes the framework of our proposed approach. To build a recommender for SPARQL query, we take the original SPARQL query and the access logs of an existing RDF repository as the input. The whole process consists of two phases, an offline process and an online process. In the offline process, we clean the original access logs through the preprocessor, then analyze the cleaned logs to find out the user access patterns. The extracted patterns can be further used to recommend queries. This phase can be done offline. In the online process, when a user sends a SPARQL query request{, the result set from the executor as well as the original query will be the input of the recommender. The recommender will take the access patterns, the original query and the size of result set into consideration, so as to generate a proper query. The recommended query will be shown to the SPARQL user.

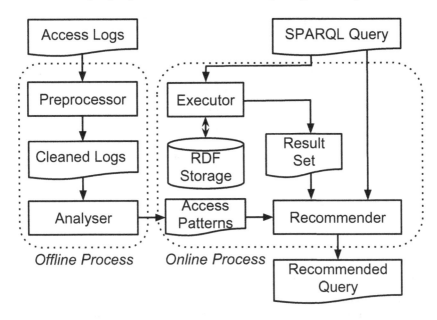

Fig. 1. The proposed framework of our approach.

The following Sect. 3 shows the details of how to analyze the access logs so as to get access patterns, while Sect. 4 describes how to build a recommender for SPARQL queries.

3   Log Analysis

In this module, we try to identify key user access patterns via log analysis, so that the recommended queries can reflect the actual user preference in real-world applications. The access logs in RDF repositories can be classified into web logs and SPARQL logs. Details of a typical web log are shown in Table 1, and it's also the standard format supported by various web servers. Such web logs are usually generated while users are browsing the RDF graph from one resource to another resource. To recommend SPARQL queries through utilizing the web logs, a comprehensive recommender should cover the following aspects:

Subject Access Pattern. A typical SPARQL query in exploring RDF graphs usually visits one or more subjects (usually URIs) in the RDF graph. And in some cases, the objects (usually URIs) of the query are also the subjects of other triples. In the real-world workload, different subjects or subject categories have diverse probabilities to be accessed.

Predicate Access Pattern. In RDF graphs, predicates with different labels reflect different relations between subjects and objects. When users explores the RDF graph, these subjects and objects are specific UIRs in most cases.

Table 1. A typical format of web log.

Entity name	Construct
Ip Address	IP of the client or visitor which made the request to the server
Time	When the server finished processing the request
Request Line	It contains the method (i.e., GET, POST etc.), the resource and the protocol of client
Status Code	It reveals whether the request resulted in a successful response, a redirection, an error caused by the client, or an error in the server
Referred Site	What the client reports having been referred from
User Agent	The identified information of browser or client

Hence, different predicates usually have different access patterns dependent on the relation types and semantics.

Subject Similarity. In a short time, namely a conversation session, a user may access multiple subjects, among which may have rich associations. Some subjects may be actually similar in some way, for example, they always exist in the same session together or there may be a predicate between them in some triples.

We can identify the above user access patterns by analyzing the web logs shown in Table 1. About the subject access distribution, a straightforward way is to directly count the frequency of a specific URI in the request line of web log. But for a large dataset, only part of URIs are accessed and appear in the web logs. The straightforward way cannot cover those URIs that have not existed in the logs before. Hence we firstly estimate the probability of each category in logs. Then we figure out the probability of each subject in the corresponding category, for example, a normal distribution or a uniform distribution which can either be defined by users or estimated in the logs. For the predicate access distribution, there's no need to consider all kinds of predicates in RDF graphs. We mainly focus on those predicates that connect multiple URIs. We find out the predicates among different subjects in a conversation session so as to estimate the probability of each kind of predicate labels. As to calculate the subject similarity, there are lots of existing methods which have been well researched in the recommendation system field. We can adopt those methods in log analysis.

A SPARQL log is similar to the web log, only the request line records SPARQL queries instead of resource URIs. In our collected logs, there are few SPARQL logs. Therefore, we only utilize web logs to make recommendation.

4   SPARQL Query Recommendation

In this section, we describe details of SPARQL query recommendation. We represent some basic recommendation methods, then explain our algorithm for recommendation.

Here we simply introduce popularity based and similarity based recommendation methods. Popularity based recommendation is a simple but efficient approach. Rather than selecting items randomly, the method ranks items according to their popularity. Item popularity is computed based on how many times the item has been accessed. Similarity based recommendation starts with the hypothesis that similar users may be interested in similar items. Therefore we could recommend subjects which the user's similar visitors have queried before or which are similar to other subjects the user himself has queried before. The similarity of subjects is shown as follows:

$$Sim(s_1, s_2) = \begin{cases} \frac{N(Sess(s_1) \cap Sess(s_2))}{N(Sess(s_1) \cup Sess(s_2))}, & Sess(s_1) \not\subset \emptyset \ \& \ Sess(s_2) \not\subset \emptyset \\ 0, & Sess(s_1) \subset \emptyset \ || \ Sess(s_2) \subset \emptyset \end{cases} \quad (1)$$

$Sim(s_1, s_2)$ is the similarity between subject s_1 and subject s_2. $Sess(s)$ represents all sessions that contain the subject s, and $N(session)$ is the number of occurrence of $session$. Note that, Eq. 1 only defines how to calculate subject similarity through web log analysis. We can also estimate similarity according to the RDF graph when similarity information is missing in the web logs, and the details can refer to the description of Algorithm 3.

Analogously, the similarity between users can be defined as Eq. 2.

$$Sim_(u_1, u_2) = \begin{cases} \frac{N(Sub(u_1) \cap Sub(u_2))}{N(Sub(u_1) \cup Sub(u_2))}, & Sub(u_1) \not\subset \emptyset \ \& \ Sub(u_2) \not\subset \emptyset \\ 0, & Sub(u_1) \subset \emptyset \ || \ Sub(u_2) \subset \emptyset \end{cases} \quad (2)$$

$Sim(u_1, u_2)$ is the similarity between user u_1 and user u_2. $Sub(u)$ represents all subjects that exist in queries of user u.

Algorithm 1. SPARQL Query Recommendation.

1: **function** QueryRecommender (q, t_U, t_L, $P_S()$, $P_P()$, $Sim()$)
2:   //Input: A SPARQL Query q, Upper Bound and Lower Bound of the Threshold
    t_U, t_L, Subject Access Distribution $P_S()$, Predicate Access Distribution $P_P()$ and
    Similarity of Subjects $Sim()$.
3:   //Output: A Recommended Query q'.
4:   $m \leftarrow ResultSize(q)$
5:   **if** $m > t_U$ **then**
6:     $q' \leftarrow Filter(q, P_S(), P_P())$
7:   **else if** $m < t_L$ **then**
8:     $q' \leftarrow Union(q, Sim())$
9:   **else**
10:    $q' \leftarrow q$
11:   **end if**
12:   **return** q'
13: **end function**

We utilize popularity and similarity to help SPARQL query recommendation, as shown in Algorithm 1. For a query q and bounds of result set threshold

t_U, t_L, we estimate the size m of the result set of original query q. If the result set is larger than the upper bound t_U, then call the $Filter(q)$ function to reduce some returned results based on the subject access distribution ($P_S()$) as well as predicate access distribution ($P_P()$). Otherwise, call the $Union(q)$ function to bind subject according to similarity.

Algorithm 2. Filter Function in SPARQL Query Recommendation.

```
 1: function Filter (q, P_S(), P_P())
 2:     //Input: A SPARQL Query q, Subject Access Distribution P_S(), Predicate
        Access Distribution P_P().
 3:     //Output: A Recommended Query q' with Filter Conditions.
 4:     beRecom ← false
 5:     for each subject s ∈ q do
 6:         if s is unbound then
 7:             c ← PopCategory(P_S())
 8:             s' ← PopSubject(c)
 9:             q' ← rewrite q with filter:filter (?s =< s' >)
10:             beRecom ← true
11:             break
12:         end if
13:     end for
14:     if beRecom = false then
15:         for each predicate p ∈ q do
16:             if p is unbound then
17:                 p' ← PopPredicate(P_P())
18:                 q' ← rewrite q with filter:filter (?p =< p' >)
19:                 beRecom ← true
20:                 break
21:             end if
22:         end for
23:     else
24:         q' ← q
25:     end if
26:     return q'
27: end function
```

As shown in Algorithm 2, in function $Filter(p)$, for the first unbound subject s, we randomly select a popular subject s' according to subject access distribution $P_S()$ and add a filter clause to bind s with p. If all subjects are bound, we randomly select a popular predicate p based on predicate access distribution $P_P()$ and approximately add a filter to the first unbound predicate. And if all subjects as well as predicates in query s are bound, we believe that the user is sure of what he wants to query with high probability so that it may be unnecessary to recommend any other query to him.

We would like to emphasize that in the selection of subject (line 7 and line 8 in Algorithm 2), we do not directly select the most popular subject in the

RDF graph, which may result in recommending various user the same subject every time. It may be nasty if we display the same thing to the same user all the time. We randomly select a "popular" category based on the prior probability of categories, then randomly pick up a subject according to the corresponding probability. This way can increase the diversity of recommended result sets and satisfy users' various requirements.

Algorithm 3. Union Function in SPARQL Query Recommendation.

1: **function** Union (q, $Sim()$)
2:     //Input: A SPARQL Query q, Similarity of Subjects $Sim()$.
3:     //Output: A Recommended Query q' with Union Conditions.
4:     $beRecom \leftarrow false$
5:     **for** each subject $s \in q$ **do**
6:         **if** s is bound **then**
7:             $s' \leftarrow$ MostSimilar($Sim()$)
8:             $q' \leftarrow$ rewrite q with union:union $q(s')$
9:             $beRecom \leftarrow true$
10:            **break**
11:        **end if**
12:    **end for**
13:    **if** $beRecom = false$ **then**
14:        $q' \leftarrow q$
15:    **end if**
16:    **return** q'
17: **end function**

As shown in Algorithm 3, in $Union()$ function, for the first bound subject s, we find the most similar s' based on $Sim()$ and add a clause $q(s')$ to union it with the original query q. If all subject are unbound, do nothing but just return the original query q. Note that if $Sim(s, *)$ does not exist in the web logs (according to Eq. 1, the value of $Sim(s, *)$ will be 0), we can directly define it based on predicates between s and $*$ in the RDF graph. For example, we can define that subject s_1 and object s_2 is similar if there exists a triple $\{s_1, owl: SameAs, s_2\}$ in the RDF graph. Furthermore, in this algorithm frame, we can union queries of similar users to the original query q if we can identify different users in the web logs. And many stat-of-art recommendation methods, such as collaborative filtering, can be used in this phase. As these methods have been well researched, we can directly adopt them.

5   Implementation and Results

5.1   Zhishi.me Dataset

In this paper, we implement the proposed method based on the Zhishi.me repository [21], one of the largest Chinese Linking Open Data. Zhishi.me

Table 2. Overall statistics of Zhishi.me.

Items	Number
Subjects/Objects	69,899,890
Predicates	149,560,207
Web Logs	10,287,178
SPARQL Logs	42,684

contains resources (individuals and categories) from three major Chinese ency-clopedia websites, namely Baidu Baike[1], Hudong Baike[2] and Chinese Wikipedia[3]. Zhishi.me started to provide services since November, 2012 and has been running for over a year. Table 2 shows statistics on the Zhishi.me dataset. We list the number of subjects/objects as well as predicates. The logs contains large amount of records. Most SPARQL logs are in simple "SPO" form, hence we mainly focus on web log analysis.

5.2   Access Patterns of Zhishi.me

As discussed in Sect. 2, to recommend query that reflects user access patterns and help user to more easily explore Zhishi.me, we analyze the web logs of Zhishi.me.

Figure 2 shows a specific example of Zhishi.me web logs and all essential entities are highlighted. Before analyzing the web logs, it's important to clean the web logs through the preprocessor (as shown in Fig. 1). We eliminate use-less records, including spider/crawler records, static file records as well as status error records. Those log records have the following key characteristic. For spi-der/crawler records, their user agents usually contain keywords like "Googlebot" or "Baiduspider". The request lines of static file records have words like "/sta-tic/*". Those records which contain error status code should also be removed.

From the cleaned web logs, we found most of the access records were gen-erated by the keyword lookup function of Zhishi.me. In look up service, one

Fig. 2. An example of Zhishi.me web log.

---

[1] http://baike.baidu.com/

[2] http://www.baike.com/

[3] http://zh.wikipedia.org/

Table 3. Numbers of resource visit and keyword lookup in the top categories.

Top Categories	Visits	Keywords
History	2756	1016
Nature	1785	711
Culture	2591	965
Technology	862	486
People	4205	2431
Science	4505	2000
Art	1267	698
Geography	5502	1811
Life	2572	1039
Society	3094	1460
Economy	13	2
Sport	742	227
Hot	1857	895

keyword may belong to multiple categories. Table 3 shows the details of resource visit and keyword lookup in 13 top categories. From the table, the category *Geography* has most visits while the category *People* has more various keywords than any other categories. The lookup frequency of all the keywords appeared in the web logs meets the power law distribution. The keywords that were queried once or twice occupy more than 87 %, while those that occurred more than 100 times in the web logs are only 11 different keywords in total. This shows that the Zhishi.me resources has a long-tail visit, and it's why we turn to Zhishi.me categories for subject recommendation.

Zhishi.me lacks of visitor's personal information and we cannot identify users through web logs only, hence we only apply the subject similarity in our method. The similarity is related to specific subject pairs, which we cannot list all here. For those subjects that do not exist in the web logs, we estimate their similarity according to the RDF graph, for example, we define their $owl:SameAs$ objects, $InternalLink$ objects as their most and second most similar subjects respectively.

5.3 SPARQL Recommendation Result

In this section, we present examples of SPARQL query recommendation based on Algorithm 1 as well as query results before and after recommendation. We take the "SPO" SPARQL query as an example, which is not only widely used in RDF repository exploration scenarios but also easily adapted to more complex and comprehensive queries. There are no ground truths available so that we have to do manual evaluation. The upper bound t_U is set to 100 and the lower bound t_L is set to 10 here. The two examples are shown in Table 4.

Table 4. Examples of SPARQL query recommendation in Zhishi.me.

Item	Original SPARQL Query	Recommended Query

q_1	select ?s ?p ?o
	where {
	  ?s ?p ?o
	}
		select ?s ?p ?o
		where {
		  ?s ?p ?o. **filter** (?s =
		  <http://zhishi.me/zhwiki/resource/
		  China>)
		}

	Results Referred to Figure 3	Referred to Figure 4

q_2	select ?s ?p ?o
	where {
	  <http://zhishi.me/zhwiki/
	  resource/Peking> ?p ?o
	}
		select ?s ?p ?o
		where {
		  {<http://zhishi.me/zhwiki/resource/
		  Peking> ?p ?o}
		  union
		  {<http://zhishi.me/zhwiki/resource/
		  %E5%8C%97%E4%BA%AC%E5
		  %B8%82> ?p ?o}
		}

	Results Referred to Figure 5	Referred to Figure 6

An example with filter function. Take q_1 in Table 4 as an example. When a new user explores a unfamiliar RDF repository, he may have no idea about what he looks for at first. He may have to send out a "SPO" SPARQL query without any specific subjects or predicates. Usually he gets lots of random and meaningless triples, just show in Fig. 3. In the recommendation stage, we recommend the subject "China" and bind it to his original query, due to the popularity probability of "China" in web logs, and the recommended query is shown in Table 4. Figure 4 shows the result after recommendation, which is more readable and helps to explore Zhishi.me better with SPARQL queries.

An example with union function. When a fresh user browses Zhishi.me with specific purpose, for instance, he's looking for "Peking", then he sends a SPARQL query like q_2 in Table 4. But unfortunately, as Fig. 5 shows, there are limited triples about "Peking" in Zhishi.me, which may not meet his satisfaction. If we recommend him the most similar subject (means "Beijing City" in Chinese and is encoded as "%E5%8C%97%E4%BA%AC%E5%B8%82") related to "Peking", and union the recommended subject to his original query, he now gets more related triples, as shown in Fig. 6.

Although we just show some simple examples here, we believe that users will benefit from proper recommendation. There is no necessary to brutally replace the original result with recommended results. Instead, we can show both the recommended result and the original one, so users can easily compare them, getting benefits from recommendation.

s	p	o
8697397fc0aca33028388a99.jpg	rights	0$8697397fc0aca33028388a99
8697397fc0aca33028388a99.jpg	rdfs: label	校园风光(图3)
8697397fc0aca33028388a99.jpg	foaf: thumbnail	8697397fc0aca33028388a99.jpg
baidu: 赵永莱	zhishi: externalLink	zbb101a_wf@@@@@@@@@@@
baidu: 赵永莱	rdfs: label	赵永莱
e6508eef808327dace1b3ed8.jpg	rights	0$e6508eef808327dace1b3ed8
e6508eef808327dace1b3ed8.jpg	rdfs: label	ryan
e6508eef808327dace1b3ed8.jpg	foaf: thumbnail	e6508eef808327dace1b3ed8.jpg
baidu: 威海绿茶	zhishi: relatedImage	f15e2429c8e87dc499250afd.jpg
32bb9c8bc76893bafc1f1044.jpg	rights	0$32bb9c8bc76893bafc1f1044

Fig. 3. Results of q_1 before recommendation.

s	p	o
zhwiki: China	zhishi: revisionID	488176
zhwiki: China	foaf: page	China
zhwiki: China	zhishi: resourceID	1139325
zhwiki: China	zhishi: pageRedirects	zhwiki: 中国
zhwiki: China	rdfs: label	China

Fig. 4. Results of q_1 after recommendation.

s	p	o
	foaf: page	Peking
	zhishi: resourceID	365525
	rdfs: label	Peking
	zhishi: revisionID	87
	sameAs	hudong: 北京
	sameAs	baidu: 北京
	zhishi: internalLink	zhwiki: 北京市

Fig. 5. Results of q_2 before recommendation.

s	p	o
	foaf: page	Peking
	zhishi: resourceID	365525
	rdfs: label	Peking
	zhishi: revisionID	87
	sameAs	hudong: 北京
	sameAs	baidu: 北京
	zhishi: internalLink	zhwiki: 北京市
	zhishi: revisionID	549962
	zhwiki: 时区	UTC+8（东八区）
	zhwiki: 气候	暖温带半湿润大陆性季风气候
	zhwiki: 中华人民共和国境内地区邮政编码列表	100000

Fig. 6. Results of q_2 after recommendation.

6   Related Work

Nowadays, RDF, the W3C recommended standard for Web data interchange, has been well studied for its storage and query answering [20,25]. A mass of RDF repositories have been designed and developed to store more than billions of RDF triples [6,18,19]. The public RDF graphs make up of the Linking Open Data cloud [5], which have been stored in a variety of RDF repositories. They are different from each other. For instance, Bio2RDF [4] consists of RDF graphs on bioinformatics knowledge. DBPedia [2], one of the largest interlinking hubs on the LOD cloud, stores encyclopedia knowledge in 119 languages. What's more, some public RDF repositories on the cloud focus on collecting language-specific knowledge, such as Zhishi.me [21], one of the largest Chinese LOD repository, which contains millions of entities extracted from three Chinese encyclopaedia websites (i.e., Chinese Wikipedia, Baidu Baike and Hudong Baike) and provides SPARQL query as well as keyword lookup services.

To help user traverse RDF graphs more easily, recommendation methods could be of great help. Recommendation techniques aim to provide suggestions for items to be of use to a user [10,23], which are related to various decision-making processes, such as what item to buy, or which SPARQL query to send in this paper. "Item" here is a general term that denotes what's recommended to the user. The basic recommendation approaches include collaborative filtering, content-based and knowledge-based ones [17]. The collaborative filtering recommends to the user the items that other user with similar tastes liked before [7,27]. It's considered as the most popular and widely implemented technique in recommendation systems. There are two primary approaches in the field of collaborative filtering: memory-based approaches [26] and model-based approaches [15]. Memory-based approaches are based on similarity among users and items, while model-based approaches use a latent factor model to capture users' preferences. The content-based methods learn to recommend items that are similar to the ones the user liked before [3,22]. The similarity of items is calculated according to the features associated with the compared items. Knowledge-based approaches recommend items based on specific domain knowledge about how certain item features meet users' needs and preferences as well as how the item is useful for the user. These methods are case-based [8,24]. All the approaches mentioned above are of great value to make SPARQL query recommendation.

To make a reasonable SPARQL query recommendation, we need to analyze the web log so as to extract user preference. The interaction between the user and RDF repositories can be analyzed and studied to gather user preferences and to "learn" what the user likes the most [1]. Query suggestion [12,28], expansion [13,14] as well as classification [11,16] in web log analysis have attracted years of research efforts.

7   Conclusion and Future Work

In this paper, we propose a framework of SPARQL query recommendation to help users easily explore RDF repositories with the SPARQL endpoint.

We extracts the users' preference through analyzing web logs of the lookup service in existing RDF repositories. The extracted user access patterns, as well as the original SPARQL query and results are used to recommend a new SPARQL query. Finally, we finish our implementation of our method on Zhishi.me, one of the largest Chinese Linking Open Data with about 150 million triples as well as over one-year web logs. We believe that the proposed approach will further help the SPARQL query research.

We also realize that there is still plenty of room to improve our implementation. For example, although user's login information is missing in Zhishi.me, it's possible to roughly distinguish individuals by utilizing ip address, time as well as user agents in the web logs, so that we can apply more complex and advanced recommendation techniques to get better recommendation results. We will keep working on it in the future.

References

1. Agosti, M., Crivellari, F., Di Nunzio, G.M.: Web log analysis: a review of a decade of studies about information acquisition, inspection and interpretation of user interaction. Data Min. Knowl. Disc. **24**(3), 663–696 (2012)
2. Auer, S., Bizer, C., Kobilarov, G., Lehmann, J., Cyganiak, R., Ives, Z.G.: DBpedia: a nucleus for a web of open data. In: Aberer, K., et al. (eds.) ISWC/ASWC 2007. LNCS, vol. 4825, pp. 722–735. Springer, Heidelberg (2007)
3. Balabanović, M., Shoham, Y.: Fab: content-based, collaborative recommendation. Commun. ACM **40**(3), 66–72 (1997)
4. Belleau, F., Nolin, M.A., Tourigny, N., Rigault, P., Morissette, J.: Bio2RDF: towards a mashup to build bioinformatics knowledge systems. J. Biomed. Inf. **41**(5), 706–716 (2008)
5. Bizer, C., Heath, T., Idehen, K., Berners-Lee, T.: Linked data on the web (ldow2008). In: Proceedings of the 17th International Conference on World Wide Web, pp. 1265–1266. ACM (2008)
6. Bornea, M.A., Dolby, J., Kementsietsidis, A., Srinivas, K., Dantressangle, P., Udrea, O., Bhattacharjee, B.: Building an efficient RDF store over a relational database. In: Proceedings of the 2013 International Conference on Management of Data, pp. 121–132. ACM (2013)
7. Breese, J.S., Heckerman, D., Kadie, C.: Empirical analysis of predictive algorithms for collaborative filtering. In: Proceedings of the Fourteenth Conference on Uncertainty in Artificial Intelligence, pp. 43–52. Morgan Kaufmann Publishers Inc. (1998)
8. Bridge, D., Göker, M.H., McGinty, L., Smyth, B.: Case-based recommender systems. Knowl. Eng. Rev. **20**(3), 315–320 (2005)
9. Burke, R.: Knowledge-based recommender systems. Encycl. Libr. Inf. Syst. **69**(Suppl. 32), 175–186 (2000)
10. Burke, R.: Hybrid web recommender systems. In: Brusilovsky, P., Kobsa, A., Nejdl, W. (eds.) Adaptive Web 2007. LNCS, vol. 4321, pp. 377–408. Springer, Heidelberg (2007)
11. Chuang, S.L., Chien, L.F.: Enriching web taxonomies through subject categorization of query terms from search engine logs. Decis. Support Syst. **35**(1), 113–127 (2003)

12. Chuang, S.L., Pu, H.T., Lu, W.H., Chien, L.F.: Auto-construction of a live thesaurus from search term logs for interactive web search (poster session). In: Proceedings of the 23rd Annual International ACM SIGIR Conference on Research and Development in Information Retrieval, pp. 334–336. ACM (2000)
13. Cui, H., Wen, J.R., Nie, J.Y., Ma, W.Y.: Query expansion by mining user logs. IEEE Trans. Knowl. Data Eng. **15**(4), 829–839 (2003)
14. Cui, H., Wen, J.R., Nie, J.Y., Ma, W.Y.: Probabilistic query expansion using query logs. In: Proceedings of the 11th International Conference on World Wide Web, pp. 325–332. ACM (2002)
15. Hofmann, T.: Latent semantic models for collaborative filtering. ACM Trans. Inf. Syst. (TOIS) **22**(1), 89–115 (2004)
16. Huang, C.C., Chuang, S.L., Chien, L.F.: Using a web-based categorization approach to generate thematic metadata from texts. ACM Trans. Asian Lang. Inf. Process. (TALIP) **3**(3), 190–212 (2004)
17. Jannach, D., Friedrich, G.: Tutorial: Recommender systems. In: Proceedings of the International Joint Conference on Artificial Intelligence, Barcelona (2011)
18. McBride, B.: Jena: a semantic web toolkit. IEEE Internet Comput. **6**(6), 55–59 (2002)
19. Neumann, T., Weikum, G.: The RDF-3X engine for scalable management of RDF data. VLDB J. **19**(1), 91–113 (2010)
20. Nitta, K., Savnik, I.: Survey of RDF storage managers. In: DBKDA 2014, The Sixth International Conference on Advances in Databases, Knowledge, and Data Applications, pp. 148–153 (2014)
21. Niu, X., Sun, X., Wang, H., Rong, S., Qi, G., Yu, Y.: Zhishi.me - weaving chinese linking open data. In: Aroyo, L., Welty, C., Alani, H., Taylor, J., Bernstein, A., Kagal, L., Noy, N., Blomqvist, E. (eds.) ISWC 2011, Part II. LNCS, vol. 7032, pp. 205–220. Springer, Heidelberg (2011)
22. Pazzani, M.J., Billsus, D.: Content-based recommendation systems. In: Brusilovsky, P., Kobsa, A., Nejdl, W. (eds.) Adaptive Web 2007. LNCS, vol. 4321, pp. 325–341. Springer, Heidelberg (2007)
23. Resnick, P., Varian, H.R.: Recommender systems. Commun. ACM **40**(3), 56–58 (1997)
24. Ricci, F., Cavada, D., Mirzadeh, N., Venturini, A.: Case-based travel recommendations. In: Wober, K.W., Frew, A., Hitz, M. (eds.) Destination Recommendation Systems: Behavioural Foundations and Applications, pp. 67–93. CABI Publishing, Wallingford (2006)
25. Sakr, S., Al-Naymat, G.: Relational processing of RDF queries: a survey. SIGMOD Rec. **38**(4), 23–28 (2010). http://doi.acm.org/10.1145/1815948.1815953
26. Sarwar, B., Karypis, G., Konstan, J., Riedl, J.: Item-based collaborative filtering recommendation algorithms. In: Proceedings of the 10th International Conference on World Wide Web, pp. 285–295. ACM (2001)
27. Schafer, J.B., Frankowski, D., Herlocker, J., Sen, S.: Collaborative filtering recommender systems. In: Brusilovsky, P., Kobsa, A., Nejdl, W. (eds.) Adaptive Web 2007. LNCS, vol. 4321, pp. 291–324. Springer, Heidelberg (2007)
28. Zhang, Z., Nasraoui, O.: Mining search engine query logs for query recommendation. In: Proceedings of the 15th International Conference on World Wide Web, pp. 1039–1040. ACM (2006)

A Semi-supervised Learning Approach for Ontology Matching

Zhichun Wang[(✉)]

College of Information Science and Technology, Beijing Normal University,
Beijing, China
zcwang@bnu.edu.cn

Abstract. Ontology matching is the task of finding correspondences between semantically related entities in different ontologies, which is a key solution to the semantic heterogeneity problem. Recently, several supervised learning approaches for ontology matching have been proposed, which outperform traditional unsupervised approaches. The existing learning based approaches treat the similarity values of matchers as normal numerical features, and need a lot of training examples. In this paper, we propose a semi-supervised learning approach for ontology matching. Our approach needs a small set of training examples, and exploit the dominant relation of similarity metrics to enrich the training examples. A label propagation algorithm is used to determine the matching results. Experimental results show that our approach can achieve good matching results with a few training examples.

Keywords: Ontology matching · Semi-supervised learning · Heterogeneity

1 Introduction

An ontology provides a vocabulary describing a domain of interest, it plays an important role in sharing and reusing knowledge among software agents. Recently, a large amount of ontologies have been created by different individuals and organizations, especially in the domain of Biology and Semantic Web. Ontologies within the same specific domain could be heterogeneous; for example, entities with the same meaning may have different names and descriptions, entities having the same name may also have different meanings.

Ontology matching is the task of finding correspondences between semantically related entities of different ontologies [8], which is critical important for solving the semantic heterogeneity problem. Currently, many ontology matching approaches have been proposed in recent years. Typically, an ontology matching approach first uses several matchers (usually utilize different types of information, such as entities' labels, entities' instances, ontology taxonomy structures) to compute the similarity between ontology entities, and then makes decision based on similarity values of entity pairs. A simple strategy is to sum the similarity

© Springer-Verlag Berlin Heidelberg 2014
D. Zhao et al. (Eds.): CSWS 2014, CCIS 480, pp. 17–28, 2014.
DOI: 10.1007/978-3-662-45495-4_2

values of each entity pair in a linear weighted fashion, and select a proper threshold to distinguish matching and non-matching pairs. However, given a matching situation, it is difficult to determine the right weights for each matcher. Some sophisticated weighting strategy such as Harmony [14], and Local Confidence [11] are proposed to adaptively determine the weights. But there is no single strategy always return the best results according to experimental results in [15].

Finding matching pairs in completely unsupervised way can not always get desired results, hence, several machine learning based ontology matching methods have been proposed to get more accurate matching results. Using a set of validated matching pairs as training examples, matching learning based methods train regression [3] or classification [7] models to infer the right matches from all the candidate matching pairs. Comparing with unsupervised approaches, machine learning based approaches usually get better results. However, we have found that two aspects in these approaches can be further improved: first, training predicting models needs a certain amount of training examples, which are not always available; second, existing methods only treat the similarity values merely as numeric features, without taking their essential characteristics into account.

In this paper, we propose a semi-supervised learning approach for ontology matching. Given a small set of validated matching entity pairs, our approach first exploit the dominant relations in the similarity space to enrich positive training examples. After getting more training examples, a graph based semi-supervised learning algorithm, label propagation [19], is employed to classify the rest candidate entity pairs into matched and non-matched groups. We also define several constraints to adjust the probability matrix in label propagation algorithm, which help to improve the quality of matching results.

This paper is structured as follows. First, we introduce the ontology matching problem and the label propagation algorithm. Then we present the method of training examples enrichment. After that we introduce the implementation of constrained semi-supervised learning algorithm for ontology matching. Finally, we evaluate the proposed algorithm on matching tests from OAEI[1].

2   Ontology Matching

An ontology is a formal specification of a shared conceptualization, which provides a vocabulary describing a domain of interest [8]. Different from schemas in the database domain, ontologies contain more semantic metadata making the entities in ontologies self-describing. The major components of an ontology are classes, properties, instances and axioms. Classes represent sets of entities or 'things' within a domain; they can be organized into a hierarchy. Properties describe various features and attributes of concepts and constraints on these attributes, which can also be organized into a hierarchy. Instances are 'things'

---

[1] http://oaei.ontologymatching.org/

represented by the concepts. Axioms are assertions in the form of logic to constrain values for classes or properties. Formally an ontology is represented as a 6-tuple [13]:

$$O = \{C, P, H^C, H^P, I, A^O\} \tag{1}$$

where C and P denote classes and properties, H^C and H^P are the hierarchy of classes and properties, I is a set of instances and A^O is a set of axioms. Ontologies can be described by several standard languages, such as the Web Ontology Language (OWL) and Resource Description Framework Schema (RDFS).

The goal of ontology matching is to find correspondences between semantically related entities in different ontologies. In this paper, we only focus on finding the equivalent relations. Here we use the term of 'entity' to refer class, property or instance in an ontology. Given a source ontology O_S, a target ontology O_T, and a entity e_i in O_S, the procedure to find the semantically equivalent entity e_j in O_T is called ontology matching, denoted as M. Formally, for each entity $e_i \in O_S$, the ontology matching M can be represented as [16]:

$$M(e_i, O_S, O_T) = e_j \tag{2}$$

3   Semi-supervised Learning for Ontology Matching

In this paper, we formulate the ontology matching problem as a binary classification problem. For each candidate matching, we first use multiple matchers to compute several similarities between the entity pair, and use these similarities as the features of candidate matching. Then we use a semi-supervised learning algorithm, label propagation, to classify candidate matchings to correct matchings and incorrect matchings. We present the details of our algorithm in this section.

3.1   Similarity Metrics and Matchers

There are a lot of ontology information of entities on which comparison can be made to compute similarities, such as label, comment and instance; [8] presents a detailed list of them. In order to compute the similarity, edit-distance and vector-based similarity are two popular similar metrics for ontology matching:

Edit-distance: Edit distance between two strings is the minimal cost of operations (insertion, replacement and deletion of characters) to be applied to one of the strings in order to obtain the other one. Given two strings s and t, the edit-distance-based similarity between them is defined as:

$$S_e(s, t) = 1 - \frac{|\{ops\}|}{\max(|s|, |t|)} \tag{3}$$

where $|\{ops\}|$ indicates the number of operations, $|s|$ and $|t|$ represent the number of characters in s and t respectively.

Vector-based similarity: Vector-based similarity is calculated between two documents in the Vector Space Model (VSM) using the TF/IDF technique. Before computing the similarity, documents are first transformed into a weighted feature vector, where the elements in the vector are weights assigned to words by TF/IDF. For a word i in document j, the weight of the word is computed as

$$\omega_{ij} = tf_{ij} \cdot \lg \frac{N}{df_i} \qquad (4)$$

where tf_{ij} is the number of occurrences of i in j, df_i is the number of documents containing i, N is the total number of documents. Then the similarity of two documents d and k is computed as the cosine value between their feature vectors:

$$S_v(d, k) = \frac{\sum_{i=1}^{M} \omega_{id} \cdot \omega_{ik}}{\sqrt{\sum_{i=1}^{M} \omega_{id}^2} \cdot \sqrt{\sum_{k=1}^{M} \omega_{ik}^2}} \qquad (5)$$

where M is total number of distinct words in all documents.

Based on the above two similarity metrics, we construct the following four matchers in our proposed approach.

(1) Name-based strategy

$$M_{name}(e_1, e_2) = S_e(label(e_1), label(e_2)) \qquad (6)$$

where $label(e)$ is the value of rdfs:label of an entity, if there is no information for rdfs:label, $label(e)$ corresponds to the last segment of the ontology entity's URI.

(2) Metadata-based strategy

$$M_{meta}(e_1, e_2) = S_v(meta(e_1), meta(e_2)) \qquad (7)$$

where $meta(e)$ is a set of words composed by combining the values of rdfs:label and rdfs:comment of entity e.

(3) Instance-based strategy

$$M_{inst}(e_1, e_2) = S_v(inst(e_1), inst(e_2)) \qquad (8)$$

If e is a class, $inst(e)$ is all metadata of the instances belonging to class e; if e is a property, $inst(e)$ corresponds to all lexical values of property e.

(4) Attribute-based strategy

$$M_{att}(e_1, e_2) = S_v(att(e_1), att(e_2)) \qquad (9)$$

where $att(e)$ denotes the values of data type properties of e.

3.2   Label Propagation

Label propagation is an effective graph based semi-supervised learning algorithm. The basic idea behind label propagation is to first build a graph in which each node represents a data point and each edge is assigned a weight usually computed as the similarity between data points, then propagate the class labels of labeled data to neighbors in the built graph in order to make predictions [17].

Let $L = \{(x_1, y_1), ..., (x_l, y_l)\}$ be the labeled data, $y \in \{1, ..., C\}$, and $U = \{x_{l+1}, ..., x_{l+u}\}$ the unlabeled data, let $n = l + u$. The basic label propagation algorithm builds a graph with the following weights:

$$\omega_{ij} = exp(-\frac{\|x_i - x_j\|^2}{\alpha^2}) \tag{10}$$

where α is a bandwidth hyper-parameter. Then it computes a $n \times n$ probabilistic transition matrix P

$$P_{ij} = P(i \to j) = \frac{\omega_{ij}}{\sum_{k=1}^{n} \omega_{ik}} \tag{11}$$

where P_{ij} is the probability of transit from node i to j. Let Y_L be a $l \times C$ label matrix which represents the labels of the labeled data, and $f = (f_L, f_U)^T$ be a $n \times C$ matrix represents the soft labels for all the nodes. The label propagation algorithm runs the following process repeatedly until convergence [18]:

- Propagate $f \leftarrow Pf$
- Clamp the labeled data $f_L = Y_L$
- Repeat from step 1 until f converges

Finally, unlabeled data $x_i \in U$ is classified to class $c \in C$ if $f_{ic} > \delta$, where δ is a threshold usually can be set to 0.5.

3.3 Matching Decision by Label Propagation

Give two ontologies O_1 and O_2, a small set of training matchings, our algorithm first predicts some matchings based on the monotonicity of similarities; then it adds the predicted matchings to the training matchings and feeds these matchings to the label propagation algorithm, which decides the rest matchings.

Enriching Training Examples

Definition 1. *Given two entity pairs $\langle e_1, e_2 \rangle$ and $\langle g_1, g_2 \rangle$, let $m = \langle m_1, m_2, ..., m_k \rangle$ and $n = \langle n_1, n_2, ..., n_k \rangle$ are similarities of $\langle e_1, e_2 \rangle$ and $\langle g_1, g_2 \rangle$ returned by k matchers, we say that $\langle e_1, e_2 \rangle$ dominates $\langle g_1, g_2 \rangle$, denote $\langle e_1, e_2 \rangle \succeq \langle g_1, g_2 \rangle$, if $m_i \geq n_i$ for all $1 \leq i \leq d$, and exist one $1 \leq j \leq d$ that $m_j \rangle n_j$.*

Definition 2. *Given two ontologies O_1 and O_2, a entity pair $\langle e, e' \rangle$, $e \in O_1$ and $e' \in O_2$, is a positive matching if e is equivalent to e', is a negative matching if e is not equivalent to e'.*

In our proposed algorithm, the user should initially verify a small set of positive matchings as training examples. Generally, more training examples enable the algorithm make more accurate predictions of the unknown matchings. Here, we make use of the dominant relations between entity pairs to infer more positive matchings. Algorithm 1 formalizes the process of enriching positive matchings, its basic idea is that the dominator of a positive matching is also a positive matching. Given a small set of positive matchings, we first use Algorithm 1 to enrich the training set, and then feed the enlarged training set to the label propagation algorithm.

Algorithm 1. Positive matchings enrichment

Input:
    Two ontologies O_1 and O_2
    A set of positive matchings E
Output:
    A new set of positive matchings $E^{'}$
1: $E^{'} \leftarrow E$
2: **for all** $\langle e, e^{'} \rangle \in E$ **do**
3:    **if** $\exists \langle g, g^{'} \rangle \succeq \langle e, e^{'} \rangle, g \in O_1, g^{'} \in O_2$ **then**
4:       $E^{'} \leftarrow E^{'} + \langle g, g^{'} \rangle$
5:    **end if**
6: **end for**

When enriching the training examples, we also add a virtual positive matching v to the training set. The similarity values of v are all set to 1. We assume that if all matchers return similarity values equal to 1 for an entity pair, then this entity pair is a positive matching.

(2)Adjust the propagation matrix

We build a kNN graph for the label propagation algorithm, in which each node only connects to the k-nearest neighbors. After obtaining the propagation matrix by Eq. 11, we adjust the propagation matrix based on some prior knowledge to make the predictions more accurate. Here, we use the following constraints (or axioms) to adjust the propagation matrix:

- An entity from O_1 can only match one entity from O_2, and vice versa;
- No crisscross matching is allowed.
- Two entities match if their *owl:sameAs* or *owl:equivalentClass, owl:equivalent-Property* entities match.

Correspondingly, we make the following adjustment to the propagation matrix:

- For each entity pair $i = \langle e, e^{'} \rangle, e \in O_1$ and $e^{'} \in O_2$, set the transition probability $P_{ij} = 0$, where $j = \langle e, g \rangle, g \in O_2/\{e^{'}\}$, or $j = \langle g, e^{'} \rangle, g \in O_1/\{e\}$.
- Let $sub(e)$ and $super(e)$ denote the sub entities and super entities of e respectively. For each entity pair $i = \langle e, e^{'} \rangle, e \in O_1$ and $e^{'} \in O_2$, set the transition probability $P_{ij} = 0$, where $j = \langle g, g^{'} \rangle$, $g \in sub(e)$ and $g^{'} \in super(e^{'})$, or $g \in super(e)$ and $g^{'} \in sub(e^{'})$.

- For two entity pairs $i = \langle e, e' \rangle$ and $j = \langle g, g' \rangle$, set the transition probability $P_{ij} = 1$ if there are *owl:sameAs* or *owl:equivalentClass*, *owl:equivalentProperty* relations between e and g, or between e' and g'.

4   Evaluation

4.1   Datasets

We evaluate our approach on four matching tasks from OAEI'2010 Benchmark dataset. These tasks are #301, #302, #303 and #304, which involve matching four real biography ontologies with one common ontology. In each matching task, there are more than 100 classes and properties in the ontologies. We evaluate our approach on these tasks because the ontologies are all realistic ones, the participants of OAEI can not get satisfying results on these tasks.

4.2   Performance Metrics

We use precision, recall, and F1-Measure to measure the performance of our proposed approach, which are defined as follows:

Precision (p): It is the percentage of correctly discovered matchings in all discovered matchings.

$$p = \frac{|A \cap T|}{|A|} \qquad (12)$$

where A is the set of discovered matchings, and T is the set of ground truth matchings.

Recall (r): It is the percentage of correctly discovered matchings in all correct matchings.

$$r = \frac{|A \cap T|}{|T|} \qquad (13)$$

Table 1. Experimental results

Tasks		#301	#302	#303	#304
0 % training	Pre.	0.90	0.96	0.90	0.92
	Rec.	0.71	0.58	0.80	0.79
5 % training	Pre.	0.91	0.96	0.90	0.95
	Rec.	0.75	0.60	0.80	0.89
10 % training	Pre.	0.93	1.00	0.90	0.96
	Rec.	0.75	0.60	0.81	0.89
15 % training	Pre.	0.94	1.00	0.91	0.96
	Rec.	0.75	0.61	0.83	0.92
20 % training	Pre.	0.96	1.00	0.91	0.97
	Rec.	0.77	0.63	0.85	0.95

F1-Measure ($F1$): F1-Measure considers the overall result of precision and recall.

$$F1 = \frac{2pr}{p+r} \tag{14}$$

4.3 Results

In the experiments, we run four group of experiments on each matching task. In these four group of experiments, we use 0 %, 5 %, 15 %, 20 % randomly selected matchings in the reference alignment as the training examples respectively. In each group of experiments, we run our algorithm ten times and each time on a different sample of training examples. When using 0 % training examples, we only use the virtual positive matching as training example.

Table 1 lists the results on four matching tasks. For each matching tasks, we list the average precision and recall on different samples of training examples.

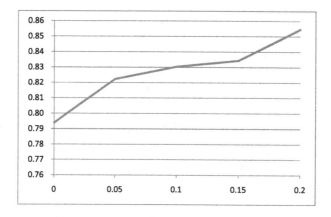

Fig. 1. F-Measure on #301 matching task

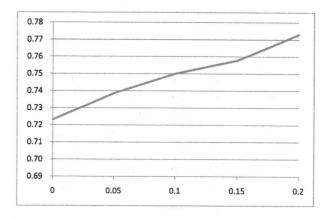

Fig. 2. F-Measure on #302 matching task

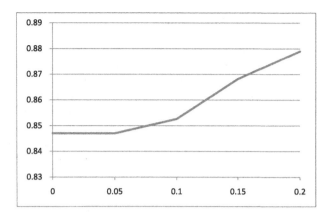

Fig. 3. F-Measure on #303 matching task

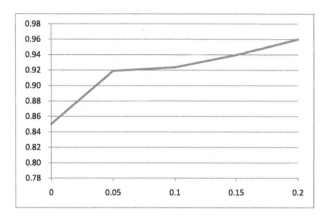

Fig. 4. F-Measure on #304 matching task

Figures 1, 2, 3 and 4 show the increments of F-measures as the number of training examples increase. It can be observed that if we use no training examples, the algorithm can also get a acceptable results. As we increase the number of training examples, the F-measures also increase. Therefore, our proposed algorithm can help improve the quality of matching results with training examples.

5 Related Work

In this section, we review related work on ontology matching.

5.1   Non-learning Based Approaches

COMA [1] is a schema matching tool, which provides a library of matching algorithms. COMA allows the user to use different algorithms and combination strategies, such as MAX, MIN, AVG and SIG, but users should select the strategies and combination method according to the matching problem. ASMOV [12] is an automated ontology matching system that combines a comprehensive set of element-level and structure-level measures of similarity with a technique that uses formal semantics to verify whether computed correspondences comply with desired characteristics. ASMOV computes four kinds of similarities, including a lexical similarity, two structural similarities and an external similarity. It combines these similarities using weighted sum, where the weights are adjusted based on static rules. PRIOR+ [14] is an adaptive ontology mapping approach, which contains three major modules, i.e. the IR-based similarity generator, the adaptive similarity filter, weighted similarity aggregator, and the neural network based constraint satisfaction solver. Both the linguistic and structural similarities of the ontologies are computed in a vector space model, and then are aggregated by an adaptive method based on their harmonies. RiMOM [13] is a dynamic multistage ontology matching framework. It uses both lexical and structural information of ontologies to compute similarity between entities. In order to adaptively combine multiple matching strategies, RiMOM estimates the similarity characteristics for each matching task by computing two factors: the label similarity factor and the structural similarity factor. UFOme [9] is a ontology matching software framework, which provides a library of matchers and a strategy prediction module which suggests individual matchers should be used in a specific task. The trategy prediction module computes three coefficients of two ontologies, including lexical affinity coefficient, structural affinity coefficient, and exploiting affinity coefficient. Based on these coefficients UFOme suggests the optimal weights value for different matchers. The strategy prediction in UFOme is quite similar to RiMOM's weighting strategy. Falcon-AO [10] is a similarity-based ontology matching system. It employs three matching strategies, i.e., V-Doc, I-Sub and GMO. V-Doc builds a virtual document for each entity, and then measures their similarity in a vector space model. I-Sub computes similarities between strings attached to different entities. GMO explores structural similarity based on a bipartite graph.

5.2   Learning Based Approaches

FOMA [5] achieves high-quality results by using a matching learning method to draw classification rules for combining results of different similarity metrics. Our approach is an unsupervised one therefore it does not need training examples. Nevertheless, it still achieves good performance in various matching tasks. GLUE [2] is a well-known machine learning based ontology matching system. It employs a set of base learners and then combine their predictions using a metalearner. Base learners apply machine learning techniques to compute the joint probability distributions of every entity pair, then use the joint distributions to compute the Jaccard similarity. The metalearner uses manually set weights to combine the base learners' predictions via the weighted sum. Eckert et al. [4] propose to use matching learning techniques to train a classifier that decides whether two elements from different ontologies should be linked by an equivalence relation based on the output of different matchers. They do not make any assumptions about the matchers, which can be simple similarity metrics or composite matching systems. The machine learning algorithm is also not particularly specified, it can be Decision Tree or Naive Bayes. APFEL [6] is another ontology matching system based on machine learning method. It computes similarities using various ontology information, and then trains a decision tree to predict the equivalence relations between the entities of two ontologies. Different from other machine learning based approaches, APFEL first uses an unsupervised approach to find some correct matchings, then uses these matchings as training examples. Most recently, [3] has proposed a ontology matching approach which interact with user feedback. Their algorithm uses logistic regression to aggregate similarities from different matchers. The training examples also are used to decide the degree of structural propagation.

6 Conclusion

In this paper, we propose a semi-supervised learning approach for ontology matching. Our approach uses four matchers to compute similarities between entities and then employs the label propagation algorithm to make the matching decisions. In order to improve the quality of matching result, we first explore the dominant relations in the similarity space to infer more training examples, and also adjust the propagation portability matrix based on domain knowledge. Experimental results show our proposed approach can effectively improve the matching results with training examples.

Acknowledgement. The work is supported by NSFC (No. 61202246), NSFC-ANR (No. 61261130588), and the Fundamental Research Funds for the Central Universities (2013NT56).

References

1. Do, H.-H., Rahm, E.: Coma: a system for flexible combination of schema matching approaches. In: Proceedings of the 28th International Conference on Very Large Data Bases (VLDB '02), pp. 610–621 (2002)
2. Doan, A., Madhavan, J., Dhamankar, R., Domingos, P., Halevy, A.: Learning to match ontologies on the semantic web. VLDB J. **12**, 303–319 (2003)
3. Duan, S., Fokoue, A., Srinivas, K.: One size does not fit all: customizing ontology alignment using user feedback. In: Patel-Schneider, P.F., Pan, Y., Hitzler, P., Mika, P., Zhang, L., Pan, J.Z., Horrocks, I., Glimm, B. (eds.) ISWC 2010, Part I. LNCS, vol. 6496, pp. 177–192. Springer, Heidelberg (2010)

4. Eckert, K., Meilicke, C., Stuckenschmidt, H.: Improving ontology matching using meta-level learning. In: Aroyo, L., Traverso, P., Ciravegna, F., Cimiano, P., Heath, T., Hyvönen, E., Mizoguchi, R., Oren, E., Sabou, M., Simperl, E. (eds.) ESWC 2009. LNCS, vol. 5554, pp. 158–172. Springer, Heidelberg (2009)
5. Ehrig, M.: Foam - framework for ontology alignment and mapping; results of the ontology alignment initiative. In: Proceedings of Integrating Ontologies Workshop (2005)
6. Ehrig, M., Staab, S., Sure, Y.: Framework for ontology alignment and mapping
7. Ehrig, M., Staab, S., Sure, Y.: Bootstrapping ontology alignment methods with apfel. In: Special Interest Tracks and Posters of the 14th International Conference on World Wide Web, WWW '05, New York, NY, USA, pp. 1148–1149. ACM (2005)
8. Euzenat, J., Shvaiko, P.: Ontology Matching, 1st edn. Springer, New York (2007)
9. Giuseppe, P., Talia, D.: Ufome: an ontology mapping system with strategy prediction capabilities. Data Knowl. Eng. **69**(5), 444–471 (2010)
10. Hu, W., Falcon-ao, Y.Q.: A practical ontology matching system. Web Semant. Sci. Serv. Agents World Wide Web **6**(3), 237–239 (2008)
11. Isabel, F.P.A., Cruz, F., Stroe, C.: Efficient selection of mappings and automatic quality-driven combination of matching methods. In: Workshop on Ontology Matching, pp. 49–60 (2009)
12. Jean-Mary, Y.R., Shironoshita, E.P., Kabuka, M.R.: Ontology matching with semantic verification. Web Semant. Sci. Serv. Agents World Wide Web **7**(3), 235–251 (2009)
13. Li, J., Tang, J., Li, Y., Luo, Q.: Rimom: a dynamic multistrategy ontology alignment framework. IEEE Trans. Knowl. Data Eng. **21**(8), 1218–1232 (2009)
14. Mao, M., Peng, Y., Spring, M.: An adaptive ontology mapping approach with neural network based constraint satisfaction. Web Semant. Sci. Serv. Agents World Wide Web **8**(1), 14–25 (2010)
15. Peukert, E., Mamann, S., Knig, K.: Comparing similarity combination methods for schema matching. In: GI Jahrestagung (1)'10, pp. 692–701 (2010)
16. Tang, J., Li, J., Liang, B., Huang, X., Li, Y., Wang, K.: Using bayesian decision for ontology mapping. Web Semant. **4**(4), 243–262 (2006)
17. Wu, M.: Label propagation for classification and ranking. Ph.D. thesis, East Lansing, MI, USA. AAI3282228 (2007)
18. Zhu, X.: Semi-supervised learning with graphs. Ph.D. thesis, Pittsburgh, PA, USA. AAI3179046 (2005)
19. Zhu, X., Ghahramani, Z., Lafferty, J.: Semi-supervised learning using Gaussian fields and harmonic functions. In:ICML, pp. 912–919 (2003)

Learning Disjointness Axioms With Association Rule Mining and Its Application to Inconsistency Detection of Linked Data

Yanfang Ma$^{(\boxtimes)}$, Huan Gao, Tianxing Wu, and Guilin Qi

School of Computer Science and Engineering, Southeast University,
Nanjing 210096, China
{myf,gh,wutianxing,gqi}@seu.edu.cn

Abstract. Disjointness between two concepts is useful to discover new information and detect inconsistencies in knowledge bases. Association rule mining, as a way to discover implicit knowledge in massive data, has been applied to learn disjointness axioms. In this paper, we first analyse the existing method to learn disjointness axioms using association rule mining. Based on the analysis, we propose an improvement of association rule mining for learning disjointness axioms. We then apply the learned disjointness axioms to inconsistency detection in DBpedia and Zhishi.me.

1 Introduction

Nowadays more and more semantic web applications are emerging continually. The validity of these applications is closely related to the quality and expressivity of linked data. Disjointness axioms, which state one concept is disjoint with another concept, are useful to discover new information and detect inconsistencies in knowledge bases. However, the expressivity of linked data is not always satisfiable: only a few knowledge bases include disjointness axioms.

Generally, the disjointness axioms at hand are created by manual annotation, which will lead to some problems. First, manual annotation is infeasiable when the knowledge base contains a large number of classes, since the relation of every two classes should be annotated. Second, manual annotation is subjective to some extent. When it comes to determine the relation of two classes, different people may hold different opinions. In this paper, we mainly discuss the acquisition of disjointness axioms by association rule mining. Association rule mining is very useful in discovering implicit knowledge. In [12], this method was used to learn schema information.

As the core dataset of the Linked Open Data Cloud, DBpedia does not contain disjointness axioms. Since association rule mining mainly takes the number of each class's instances into consideration, and the quality of "instance-of" relations in DBpedia is relatively high, it is a good choice to do experiment on DBpedia to show the validity of association rule mining. Furthermore, in order to evaluate the validity of association rule mining in noisy datasets, we also

© Springer-Verlag Berlin Heidelberg 2014
D. Zhao et al. (Eds.): CSWS 2014, CCIS 480, pp. 29–41, 2014.
DOI: 10.1007/978-3-662-45495-4_3

do experiment on Zhishi.me [9], which is the first effort to publish large-scale
Chinese linked data.

The main contributions of this paper are as follows:

- We redo the experiments of learning disjointness axioms by association rule
  mining, which has been discussed in [3], on DBpedia and Zhishi.me, and then
  compute the recall and precision compared with the gold standard we established beforehand.
- From the analysis of disjointness axioms acquired in the experiments, we identify some problems.
- According to the problems identified, an improvement of the existing method
  is proposed. We do experiment to show that the improved method can solve
  some problems and improve accuracy.
- The disjointness axioms from the experiments are applied to detect inconsistencies in DBpedia and Zhishi.me.

The rest of this paper is organized as follows: In Sect. 2 we give a brief
introduction of the related work in the field of learning disjointness axioms. In
Sect. 3, we introduce the method of applying association rule mining to learning
disjointness axioms. In Sect. 4, we do experiment on DBpedia and Zhishi.me to
evaluate the validity of association rule mining. After an analysis on the experiment results, we discuss the problems of the existing method and propose an
improvement for it. In Sect. 5, we use the disjointness axioms to detect inconsistencies in DBpedia and Zhishi.me. Section 6 summarizes the work in this paper
and provides an outlook for future research.

2   Related Work

The schema information of RDF knowledge bases may present two problems:
incorrectness and incompleteness. Faced with the huge amounts of structured or
semi-structured data, it is very difficult to manually annotate schema information.

In recent years, there have appeared some works which adopt machine learning methods to acquire disjointness axioms. In [11] the Vector Space Model has
been applied to get disjointness. All the classes are mapped to vectors in the same
dimension which is determined by the number of object properties and datatype
properties in the knowledge base. Thus the similarity of each two classes can be
computed. If the similarity value of two classes is below a given threshold, then
disjointness can be deduced. However, the time and space complexity of this disjointness learning method are relatively high, especially in datasets which contain
large amount of classes and properties. In [3,12] mainly discuss the adoption of
inductive methods into disjointness learning, include correlation computing and
association rule mining. In correlation computation, the correlation value of every
two classes should be computed, and the time cost mainly comes from SPARQL
queries. It is a relatively simple approach to get disjointness. But the recall can not
be guaranteed. As to association rule mining, the time cost mainly comes from the

multi-scanning of transaction table. Association rule mining outperforms correlation computation from the perspective of precision and recall . However, the transaction table created for mining negative association rules takes up a lot of storage space. In [12] only evaluated the validity of association rule mining by computing the precision and recall scores. Thus, the quality of the learned disjointness axioms has not been discussed. In [4] the authors introduced association rule mining into detecting deficiencies in large relational database. Transactions which contradict with the association rules must be suspected of deficiencies.

When it comes to determine the relation of two classes, the methods above mainly take the "instance-of" relations of the datasets into consideration. In order to improve the quality of disjointness learning, more features of the datasets should be considered. In [13] various features are extracted for supervised learning are taxonomic overlap, semantic similarity, etc. In this paper, 2,000 pairs of disjoint classes are used as training dataset to train a classifier. The authors not only finished an experiment on different datasets to show the validity of the learned classifier, but also performed an analysis to figure out which features contributed most to the overall performance of the classifier. ORE [5] is another case to adopt supervised learning methods for the enrichment of ontology. Currently "equivalentClass" and "subclassOf" relations can be added to the ontology.

Although there have been some works on disjointness learning, there is little work on the analysis of disjointness axioms acquired automatically. In order to improve the quality of disjointness learning, the problems exist in the present learning method should be spotted.

3   Association Rule Mining in Disjointness Learning

The basic concepts of association rule mining have been discussed in [7]. In this section, we first briefly describe the process of disjointness learning by association rule mining. Then, we do experiments on DBpedia 3.5.1 and Zhishi.me to show the validity of association rule mining. Finally, we analyze on the disjointness axioms acquired in the experiments and discuss some problems.

3.1   Method Description

For two atomic concepts A and B, $A \sqsubseteq \neg B$ means that $A \cap B = \varnothing$. That is to say, A and B are mutually disjoint, and they should have no common individuals. In disjointness learning, the task of association rule mining is to generate patterns like the form $A \rightarrow \neg B$, and then transform them to disjointness axiom "A owl:disjointWith B". Figure 1 gives the workflow of negative association rule mining. In the following, we will describe the workflow of association rule mining in detail:

- **Class acquisition.** The classes in the dataset can be acquired by a simple SPARQL query:

```
SELECT DISTINCT ?concpet WHERE
{?instance a ?concept.}
```

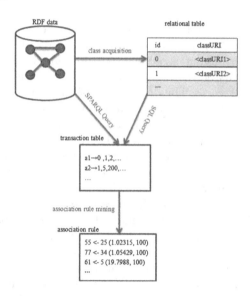

Fig. 1. Workflow of association rule mining

Classes in RDF datasets are represented by IRI. In order to simplify the experiments, we store all the classes returned by the above query in a relational table. Each class is assigned two integer identifiers to express that an instance belongs to a class(positive identifier) or not(negative identifier). Thus, in the following experiments, the items in the transaction tables are all integer numbers other than strings.

– **Transaction table creation.** For an instance i and a class C, if i is not stated to be an instance of C, we can have $i \in \neg C$, and the negative identifier will appear in the row corresponding to i otherwise the positive identifier will appear in. In the transaction table, each row contains the identifiers of all the classes an instance belongs to or not. Thus, the number of lines in the transaction table is determined by the number of instances in the dataset, while the number of integers in each row equals to the number of atomic classes in the dataset.

– **Negative association rule mining.** We use the Apriori [1] algorithm to mine association rules. Since a negative association rule involves two classes, we only need to generate frequent itemsets which contain two class.

3.2 Experiments

3.2.1 Experiment Data Setting

DBpedia [6] includes an ontology which was manually created and maintained. Besides, DBpedia community creates mappings from Wikipedia information representation structures to the DBpedia ontology. The statistical data of classes and instances in DBpedia 3.5.1 is given in Table 1.

Table 1. Statistical data from DBpedia

Total number of classes	259
Classes that have instances	240(include top class owl:Thing)
Total number of instances	1,477,796

In order to illustrate the validity of association rule mining in relatively noisy dataset, we also do experiments on Zhishi.me. However, lack of schema information makes it difficult to learn disjointness in Zhishi.me. There are no class definitions in Zhishi.me. This dataset only contains categories from the automatic extraction of Chinese online encyclopedias. In zhishi.me, there are a large number of categories but the quality of categories is poor. This is because the editors name the categories casually and without careful consideration, many categories are rarely used.

In the experiment, we manually selected 50 categories from Hudongbaike. To facilitate our experiments and ensure the reasonableness of the experiments, we should follow several criteria:

1. The selected categories should cover the 11 top categories in Fig. 2, rather than cover only a few top categories.

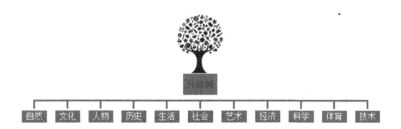

Fig. 2. Category tree of Hudongbaike

2. In Hudongbaike, the average number of instances a class has is 54. However, a considerable number of categories only have one instance. Since the number of instances a class has is a very important feature in disjointness learning, we make sure that each selected category has no less than 54 instances.
3. Each selected category corresponds to one concept in DBpedia 3.5.1. That is because, we believe that some categories in Hudongbaike should not be considered as concepts. Selecting categories according to DBpedia will also facilitate the analysis of experiment results.
4. Data noise is an obvious problem in Zhishi.me. On the one hand, some resources are used both as categories and instances, on the other hand, many child categories have more instances than their parent categories. When manually selecting, we should reserve the noisy feature of Zhishi.me rather than only choose those with relatively high quality.

5. The selected categories should better be sub-categories of several top categories in Fig. 2. For example, category "Journalist" can be sub-category of "Society", "Science", "Person" in Fig. 2. The inheritance paths are:

$$Society \longleftarrow Occupation \longleftarrow Journalist$$
$$Science \longleftarrow Sociology \longleftarrow Journalism \longleftarrow Journalist$$
$$Person \longleftarrow Sociologist \longleftarrow PersonofEveryOccupation \longleftarrow Journalist \longleftarrow MediaWorkers$$

6. The name of selected categories must be same as the name of concepts in DBpedia. In this way, we can use the gold standard to evaluate the experiment of zhishi.me Following the 6 criteria above, Table 2 lists a part of the categories we selected.

All the RDF data used in the experiments is stored in RDF triple store Sesame [2].

In order to evaluate the validity of association rule mining in disjointness learning, precision and recall should be computed. Thus, a gold standard of disjointness axioms is needed. Since there is no disjointness information in DBpedia ontology, the gold standard should be created manually. We follow the process that has been proposed in [13] to generate gold standard. The process consists of two steps: (1) Manually annotating a small part of disjointness axioms. (2) In order to get more disjointness axioms, we input the manually annotated axioms into OWL reasoner. By manually annotating and reasoning, we finally get 59,912 disjointness axioms.

3.2.2   Experiments Results

In this section, we present the experiment results of association rule mining which was conducted on DBpedia 3.5.1 and data selected from Hudongbaike.

Table 3 gives the experiment results of DBpedia when the minimum support is 0.01 %. The precision and recall are computed by comparing with the gold standard.

For the 50 categories are selected from Hudongbaike, we create a transaction table which contains 231,151 transactions and each transaction contains 50 items. The transaction table takes up a storage space of about 34.5 M. Similarly, we give the experiment results of Zhishi.me in Table 4. Unlike in DBpedia, the minimum support is set to 0.1 %.

We do not present the recall in Table 4, since the experiment data is just a small part of Hudongbaike.

3.3   Analysis on the Experiment Results

After an analysis on the disjointness axioms from association rule mining, we find some deficiencies. In the following, we mainly discuss these deficiencies.

Problem 1. Since the transaction tables are created on the basis of "instance-of" relations in the dataset, classes that have no individuals will never appear

Table 2. Part of the selected categories from Hudongbaike

Corresponding concept in DBpedia	Top category	Number of instances
Animal	Nature	19,845
Language	Culture	45,766
BasketballPlayer	Person	888
WorldHeritageSite	History	419
Song	Life	6,835
City	Society	2,364
MusicalWork	Art	17,972
Currency	Economy	378
Weapon	Science	2,436
SportsEvent	Sports	395
Spacecraft	Technology	79

in the transaction table. Consequently, they will never appear in the learned disjointness axioms.

Problem 2. We may get rules which form are "child→¬father". For example, the axiom "dbpedia:WebSite rdfs:subClassOf dbpedia:Work" from the DBpedia ontology states that each instance of class "Website" is also an instance of class "Work". In DBpedia3.5.1, "Website" has 1,852 instances. However, all these instances are not stated to be instances of "Work". By association rule mining, we get the wrong rule

$$\text{Website} \rightarrow \neg \text{Work}(\text{support} = 0.13\,\%, \text{confidence} = 100\,\%)$$

Problem 3. We may get rules which form are "father→¬child". For example,

$$\text{Person} \rightarrow \neg \text{Actor}(\text{support} = 21.5\,\%, \text{confidence} = 88\,\%)$$

Actually "Actor" is an subclass of "Person", the instances of "Person" are not necessarily instances of "Actor". Therefore, in the transaction table the combination of "Person" and "¬Actor" appeared concurrently in many transactions.

Problem 4. Among the outputted rules of Apriori, the combination of A→B and A→¬B, or A→B and B→¬A may appear at the same time. These combinations of positive rules and negative rules lead to confliction. In the experiment results of DBpedia, when set the minimum support to 0.01 %, minimum confidence to 80 %, we can get

$$\text{Person} \rightarrow \neg \text{PokerPlayer}(\text{support} = 21.5\,\%, \text{confidence} = 99.8\,\%)$$
$$\text{PokerPlayer} \rightarrow \text{Person}(\text{support} = 0.04\,\%, \text{confidence} = 100\,\%)$$

The first rule means that "Person" and "PokerPlayer" are disjoint classes, while the second one means "PokerPlayer" is a subclass of "Person". Obviously they

Table 3. Experiment results in DBpedia

Confidence	Number of negative association rules	Number of correct rules	Precision	Recall
80 %	46,551	42,153	0.704	0.906
90 %	46,521	42,153	0.704	0.906
100 %	45,915	41,843	0.698	0.911

Table 4. Experiment results in Zhishi.me

Confidence	Number of negative association rules	Number of correct rules	Precision
80 %	1,954	1,804	0.923
90 %	1,951	1,804	0.925
100 %	1,607	1,526	0.946

contradict with each other. We find that there are 280 such contradict rule pairs in the experiment results of DBpedia.

Problem 5. Rule A→¬B can be transformed to axiom "A owl:disjointWith B". However, since "owl:disjointWith" is a symmetric property, axiom "B owl:disjointWith A" also holds. Therefore, it is more reasonable that if A→¬B is outputted, B→¬A is also outputted. When we set the minimum support to 0.01 %, minimum confidence to 80 %, there are 8,507 rules like the form A→¬B which have no corresponding rules B→¬A.

The above 5 problems were found during the analysis on the experiment results of DBpedia. These problems also exist in the experiment results of Zhishi.me. Furthermore, compared with DBpedia, the quality of "instance-of" relations in Zhishi.me is not promising. On the one hand, many instances of the child category are not declared to be instances of its parent categories. On the other hand, child category may has more instances than its parent categories. Figure 3 gives an example to illustrate these two problems in Zhishi.me.

As is shown in Fig. 3, in Hudongbaike category "Hospital" has 2,822 instances, while "Organization" has 1,446 instances. These two categories have 5 common instances. However, the RDF triple "hudong:Hospital skos:broader hudong:Organization" from Zhishi.me states that "Organization" is a parent category of "Hospital". Thus, Apriori outputs the wrong rule

Hospital→¬Organization(support = 1.2 %, confidence = 99.8 %)

From the above analysis we can conclude that association rule mining closely relies on the quality of "instance-of" relations. However, in linked datasets this kind of information maybe incorrect or incomplete [10]. Thus, we may get wrong rules or mutually contradicted rules .

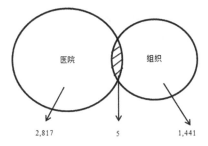

Fig. 3. Number of instances of "Hospital" and "Organization" from Hudongbaike

4   Improvement of the Present Disjointness Learning Method

From the problems discussed in Sect. 3.3, we can see that "subclass-of" relations in the ontology can be used to determine whether a negative association rule is right or not. Thus, in our improved method, we take the class hierarchy into consideration. The main idea is that for each instance, we figure out all the classes it belongs to. The improved method can be divided into two steps:

1. Get class hierarchy by association rule mining. That is to say, mine positive association rules like the form A→B, because such rules indicate that A is a subclass of B. For a class C, the purpose of this step is to get all the parent classes of C.
2. Constructing transaction table and mining negative association rules. If i is an instance of C, then all the parent classes' positive identifiers of C will appear in the transaction corresponding to i.

Since "owl:Thing" is the parent class of all classes in DBpedia, axioms that involve "owl:Thing" are excluded from the results. When we set the minimum support to 0.01 % and minimum confidence to 80 %, we get 334 positive association rules. Two of them are incorrect:

Activity→Game(support = 0.08 %, confidence = 92 %)
Infrastructure→Road(support = 0.6 %, confidence = 100 %)

Since the quality of Zhishi.me is relatively poor [8], we do not directly adopt association rule mining to get "subclass-of" relations in Zhishi.me. As mentioned before, every selected category from Hudongbaike has a corresponding class in DBpedia. Thus, we can get the hierarchy of the selected categories according to DBpedia. Figure 4 gives an example of category hierarchy in the selected data.

In Fig. 4, "Mammal" has two parent categories: "Animal" and "Species". Thus, the instances of "Mammal" are also instances of "Animal" and "Species". The experiment results of the improved method are shown in Table 5.

From Table 5 we can see that after improvement, the precision has been increased and some wrong rules are filtered out. For the 5 problems discussed

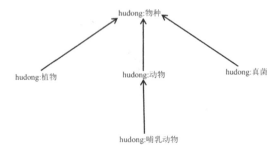

Fig. 4. A snippet of class hierarchy in Hudongbaike

Table 5. Experiment results in DBpedia

	Confidence 80 %		Confidence 90 %		Confidence 100 %	
	Total number of rules	Correct rules	Total number of rules	Correct rules	Total number of rules	Correct rules
Before improve	1,954	1,804	1,951	1,804	1,607	1,526
After improve	1,939	1,802	1,937	1,802	1,508	1,431

in Sect. 3.3, we also make a comparison between experiment results before and after improvement:

Problem 1. Since all the categories are selected from Hudongbaike have instances, there is no such problems.

Problem 2. Before improvement, Apriori outputs 10 negative association rules A→¬B, where A is a child category of B. After improvement, this problem no longer exists.

Problem 3. Before improvement, Apriori outputs 12 negative association rules A→¬B, where B is a child category of A. After improvement the number of such rules reduced to 8.

Problem 4. Before improvement, there are two pairs of contradictory rules. After improve, since more positive rules are outputted, there are 4 pairs of contradictory rules.

Problem 5. Before improvement, there are 404 rules A→¬B that have no corresponding rule B→¬A. After improvement, the number of such rules is 405.

From the analysis above, we can conclude that the improved method is quite useful to solve Problem 2, and it is helpful to solve other problems. However, it is of no use to address Problem 5.

5   Disjonitness in Inconsistency Detection

From the analysis of linked datasets, we also find logical inconsistencies which can not be detected with the present data. For example, in DBpedia we can find the following statements:

dbpedia : Nuriadbpedia : Place
dbpedia : Nuriadbpedia : Person

That is to say, resource "Nuri" is defined to be an instance of both "Place" and "Person". Obviously, "Place" and "Person" are mutually have no common individuals. Thus, the two statements above lead to logical inconsistency. However, since there are no disjointness axioms in DBpedia ontology, such inconsistencies cannot be detected. In the following, we discuss the detection of logical inconsistencies when adding disjointness axioms to the ontology.

We use SPARQL queries to detect logical inconsistencies from the DBpedia SPARQL Endpoint[1]. Table 6 gives the statistical data during the detection of logical inconsistencies.

As is shown in Table 6, the number of common individuals between disjoint class pairs is mainly in $0-10$. There are also disjoint class pairs whose common individuals exceed 10,000. Among all the disjoint classes, we find that the number of common individuals of "Infrastructure" and "Place" has reached 50,000. That is because, though we can get the axiom "Infrastructure owl:disjointWith Place" by association rule mining, the two classes are not necessarily disjoint from our own judgement. Obviously, class "Organization" and "Person" are mutually disjoint, but they have 3,333 common individuals, which leads to inconsistencies. Among the 361 pairs of disjoint classes in Table 6, 175 pairs of classes only have one common individual. For example, "Mammal" and "Place" have one common individual "Salt_Lake". From the Wikipedia page describing resource "Salt_Lake", we can see that though this resource has more than one meanings, they all have nothing to do with "Mammal".

Also, when adding the acquired disjointness axioms to Zhishi.me, we found 164 disjoint class pairs which have common individuals. Among them "World-HeritageSite" and "Architecture" have the most common individuals, which reaches 125. Besides, 66 disjoint class pairs have only one common individuals. For example, "Animal" and "Website" has one common individual: the Website "http://www.macaupanda.org.mo/", though this is a Website about animal, it is incorrect to define it as an individual of "Animal".

Table 6. Logical inconsistency detection results on DBpedia

Number of common individuals	Pairs of disjoint classes
$[1,10)$	317
$[10,100)$	27
$[100,1,000)$	7
$[1,000,10,000)$	6
$[10,000,+\infty)$	4

---

[1] https://dbpedia.org/sparql

6   Conclusion

Disjointness between atomic classes is necessary for inconsistency detection, but only a few RDF datasets contains such information. In this paper, we mainly discussed the acquisition of disjointness by association rule mining. We did experiments on DBpedia and Zhishi.me to show the validity of association rule mining in disjointness learning. We not only presented the precision and recall, but also did an analysis on quality of disjointness axioms acquired. According to the deficiencies exists in the present learning method, we proposed an improvement. Finally, the learned disjointness axioms were used to detect logical inconsistencies.

In the future, we will take more features of linked data into consideration to improve the quality of disjointness learning. New approaches to evaluate the quality of disjointness learning should be developed. Furthermore, we will try to deal with the detected inconsistencies to improve data quality.

References

1. Agrawal, R., Srikant, R., et al.: Fast algorithms for mining association rules. In: Proceedings of the 20th International Conference on Very Large Data Bases, VLDB '94, vol. 1215, pp. 487–499. Morgan Kaufmann Publishers Inc. (1994)
2. Broekstra, J., Kampman, A., van Harmelen, F.: Sesame: a generic architecture for storing and querying RDF and RDF schema. In: Horrocks, I., Hendler, J. (eds.) ISWC 2002. LNCS, vol. 2342, pp. 54–68. Springer, Heidelberg (2002)
3. Fleischhacker, D., Völker, J.: Inductive learning of disjointness axioms. In: Meersman, R., Dillon, T., Herrero, P., Kumar, A., Reichert, M., Qing, L., Ooi, B.-C., Damiani, E., Schmidt, D.C., White, J., Hauswirth, M., Hitzler, P., Mohania, M. (eds.) OTM 2011, Part II. LNCS, vol. 7045, pp. 680–697. Springer, Heidelberg (2011)
4. Hipp, J., Güntzer, U., Grimmer, U.: Data quality mining-making a virtue of necessity. In: Proceedings of the Data Mining and Knowledge Discovery Workshop, pp. 52–57 (2001)
5. Lehmann, J., Bühmann, L.: ORE - a tool for repairing and enriching knowledge bases. In: Patel-Schneider, P.F., Pan, Y., Hitzler, P., Mika, P., Zhang, L., Pan, J.Z., Horrocks, I., Glimm, B. (eds.) ISWC 2010, Part II. LNCS, vol. 6497, pp. 177–193. Springer, Heidelberg (2010)
6. Lehmann, J., Isele, R., Jakob, M., Jentzsch, A., Kontokostas, D., Mendes, P.N., Hellmann, S., Morsey, M., van Kleef, P., Auer, S., et al.: Dbpedia-a large-scale, multilingual knowledge base extracted from wikipedia. Semant. Web J. (2013)
7. Liu, B.: Web Data Mining. Springer, Heidelberg (2007)
8. Ma, Y., Qi, G.: An analysis of data quality in DBpedia and Zhishi.me. In: Qi, G., Tang, J., Du, J., Pan, J.Z., Yu, Y. (eds.) CSWS 2013. CCIS, vol. 406, pp. 106–117. Springer, Heidelberg (2013)
9. Niu, X., Sun, X., Wang, H., Rong, S., Qi, G., Yu, Y.: Zhishi.me - weaving chinese linking open data. In: Aroyo, L., Welty, C., Alani, H., Taylor, J., Bernstein, A., Kagal, L., Noy, N., Blomqvist, E. (eds.) ISWC 2011, Part II. LNCS, vol. 7032, pp. 205–220. Springer, Heidelberg (2011)

10. Paulheim, H., Bizer, C.: Type inference on noisy RDF data. In: Alani, H., Kagal, L., Fokoue, A., Groth, P., Biemann, C., Parreira, J.X., Aroyo, L., Noy, N., Welty, C., Janowicz, K. (eds.) ISWC 2013, Part I. LNCS, vol. 8218, pp. 510–525. Springer, Heidelberg (2013)

11. Töpper, G., Knuth, M., Sack, H.: Dbpedia ontology enrichment for inconsistency detection. In: Proceedings of the 8th International Conference on Semantic Systems, I-SEMANTICS '12, pp. 33–40. ACM (2012)

12. Völker, J., Niepert, M.: Statistical schema induction. In: Antoniou, G., Grobelnik, M., Simperl, E., Parsia, B., Plexousakis, D., De Leenheer, P., Pan, J. (eds.) ESWC 2011, Part I. LNCS, vol. 6643, pp. 124–138. Springer, Heidelberg (2011)

13. Völker, J., Vrandečić, D., Sure, Y., Hotho, A.: Learning disjointness. In: Franconi, E., Kifer, M., May, W. (eds.) ESWC 2007. LNCS, vol. 4519, pp. 175–189. Springer, Heidelberg (2007)

Decision Tree Learner in the Presence of Domain Knowledge

João Vieira and Cláudia Antunes[(⊠)]

Instituto Superior Técnico, Universidade de Lisboa, Lisbon, Portugal
{joao.c.vieira,claudia.antunes}@ist.utl.pt

Abstract. In the era of semantic web and big data, the need for machine learning algorithms able to exploit domain ontologies is undeniable. In the past, two divergent research lines were followed, but with background knowledge represented through domain ontologies, is now possible to develop new ontology-driven learning algorithms. In this paper, we propose a method that adds domain knowledge, represented in OWL 2, to a purely statistical decision tree learner. The new algorithm tries to find the best attributes to test in the decision tree, considering both existing attributes and new ones that can be inferred from the ontology. By exploring the set of axioms in the ontology, the algorithm is then able to determine in run-time the best level of abstraction for each attribute, infer new attributes and decide the ones to be used in the tree. Our experimental results show that our method produces smaller and more accurate trees even on data sets where all features are concrete, but specially on those where some features are only specified at higher levels of abstraction. We also show that our method performs substantially better than traditional decision tree classifiers in cases where only a small number of labeled instances are available.

Keywords: Semantic aspects of data mining · Classification · Decision trees · Background knowledge · Ontologies

1   Introduction

The automatic generation of simple and accurate classifiers from data is one of the major fields of data mining. Classification algorithms are supervised methods that look for and discover the hidden associations between the target class and the independent variables [12]. In a traditional supervised inductive learning scenario, the classifier is produced solely from a given set of already labeled instances or observations, ignoring any existing domain knowledge available.

However, with the advances of the semantic web, this knowledge is becoming increasingly available through domain ontologies, in many areas of application. Moreover, it is usual that the instances to be classified are described by features at different levels of abstraction. That is, the values of a particular attribute are specified at different levels of abstraction on different instances.

© Springer-Verlag Berlin Heidelberg 2014
D. Zhao et al. (Eds.): CSWS 2014, CCIS 480, pp. 42–55, 2014.
DOI: 10.1007/978-3-662-45495-4_4

One of the most widely known data sets in classification is the Mushroom data set [3, 11]. One might think that there is not much interest or knowledge about mushrooms beyond the circle of professional biologists or amateur witches but as it turns out there is a vibrant community of mushroom enthusiasts (and not only of the psychadelic kind) across the globe. As such, over time, there have been numerous guides and books about the subject and a great number of people have learned to correctly identify mushrooms through luck and experience, which is remarkable in an area where learning from one's mistakes might not be advisable.

One of the characteristics that are important when identifying mushrooms is its odor. Experts with gifted noses will tell you that an odor of anise or almonds is a good sign as is no odor at all. However if you smell creosote for example, it will probably be a good idea to have dinner somewhere else. But can we help amateurs with less sensitive noses avoid death by mushroom poisoning? Suppose that while some experts input an exact odor, other people can only input if it smells good or bad. One can certainly leverage existing domain knowledge to make an automated classifier aware of which odors are pleasant to humans and which are not and if possible classify the mushroom with this less precise information.

More generally if experts learn from guides and books in addition to examples and experience, why can't classifiers also make use of existing domain knowledge in addition to a set of training labeled instances?

If we start to read a book about mushrooms, we soon realize that if we want to survive a dinner of these dangerous delicacies, we will need more than knowing attribute values at different levels of abstraction. Often we are teached to infer new attributes, like species, from a combination of existing ones. For example, there is a common species of mushrooms called Parasol. Mushrooms in these species share, of course, the same basic characteristics, like cap surface and color, gill size and spacing and so on. However there is another species that share almost the same characteristics and their distinction is only possible by looking at a combination of two attributes, a bad smell and a greenish spore print color.

This is also very common in medical diagnosis, when some conditions can only be considered when certain specific combinations of symptoms occur. For example, weakness and fatigue combined with weight loss and yellow discoloration of the skin points to liver problems. It is not enough to know the cause or the treatment but it helps narrow it down. The classifier we propose can, for example, use domain knowledge to determine probable liver problems and then use that extra information and a set of labeled instances to arrive at a concrete diagnosis.

For automated classification methods to be adopted in practice, it is crucial that a relationship of trust can be established between domain experts and the models generated. When the cost of making mistakes is very high, numerical validation is usually not enough. This is why it is so important that the generated models are simple, understandable, and somewhat aligned with certain facts that are known to be true in the domain.

Beside the inability of using anything other than labeled examples, one of the major problems faced during classifiers training is the *overfitting* of the learned model to the training data. Usually resulting in excessively complicated models, with low predictive power for unseen data [6]. Overfitting is then the production of models that include more terms or use more complicated rules than necessary, compromising the fundamental rule of machine learning – the *Occam's razor* principle.

Adding irrelevant predictors can make models perform worse because the coefficients fitted to them add random variation to the subsequent predictions.

By ignoring the relationship between attribute values, e.g., the fact that *anise* and *pleasant* odors are not two unrelated features, but the same at different levels of abstraction, most current algorithms produce models that have very low portability [9].

We believe that adding domain knowledge is the key to the solution of these problems. The introduction of ontologies, as a means to formally represent existing background knowledge, in the learning process will give rise to the production of simpler and more accurate models. Since, through ontologies will be possible to consider different levels of abstraction and explore the relationship between the concepts instantiated in the data set.

In this paper, we overview the related work and propose an approach that introduces domain knowledge, represented as an ontology written in a standard format, OWL 2, in the context of classification. We propose a method that given a set of examples and an ontology expressing existing domain knowledge will automatically classify instances making use of the available domain knowledge when, and in the extent that, it helps produce simpler, more accurate decision trees and more probable classifications.

2   Literature Review

Two major approaches have dominated research in artificial intelligence: one based on logic representations, and one focused on statistical ones. The first group includes approaches like logic programming, description logics, rule induction, etc. The second, more used in machine learning, includes Bayesian networks, hidden Markov models, neural networks, etc. Logical approaches tend to focus on handling the complexity of the real world, and statistical ones the uncertainty [7] that is present in field applications.

This duality is clearly represented in classification, where a lot of efforts were taken in the last decades in the research and development of algorithms that explored certain principles of statistics to build predictive models. Examples of algorithms following this approach include SVMs [5], back-propagation [20], Naive Bayes, KNN [1], C4.5 [18], among others. These algorithms are usually very efficient in learning a model. Moreover, the models produced yield good levels of accuracy for unseen data, if training sets were properly balanced and sized. These kind of algorithms were the focus of most research in the last decades and saw wide adoption and acceptance by the industry.

On the other hand, Inductive Logic Programming (ILP) is the most known representant of the logic approach to classification [4]. In this kind of approach, in addition to the training set, an encoding of the known background knowledge is also provided. An ILP system will then derive a logic program as a hypothesis which entails all the positive and none of negative examples.

Although ILP systems benefit from relevant background knowledge to construct simple and accurate theories more quickly [21], background knowledge that contains large amounts of information that is irrelevant to the problem being considered can, and have been shown to, hinder the search for a correct explanation of the data [17]. Experimental results [8,19] also show that performance (in terms of time complexity) is much worse than statistical approaches, like C4.5. Further, traditional ILP is unable to cope with the uncertainty of real-world applications such as missing or noisy data, a known drawback when compared to the statistical approach.

Although probabilistic ILP takes a step further in terms of dealing with uncertainty it does not perform consistently better than equivalent statistical approaches in terms of accuracy. The computational complexity of the learning phase is also much higher.

There has been surprisingly little work on probabilistic learning with datasets described using formal ontologies [14]. Ontologies are crucial to deal with semantic interoperability and with heterogeneous data sets.The strengths of purely statistical methods are related with their relative simplicity and ability to work with data that underwent less preparation than required by the logic approach (where data and existing knowledge must be represented or transformed to first order logic). It is also able to scale up relatively well and is robust, i.e., performs well even when its assumptions are somewhat violated by the true model from which the data is generated. This is related with the inherent ability of the statistical approach to deal with uncertainty.

However, it ignores the complexities of the real world. First, it is not possible to express or make use of existing domain knowledge, to explicitate relationships between attributes or hierarchies of features. And second, it doesn't allow for constraining the results according to facts which are known to be true, even if not represented in the subset of data being fed to the learning algorithm.

EG2 [15] was one of first approaches to extend a purely statistical method, the ID3 decision tree learner, to make use of background knowledge in order to reduce the *classification cost.*

More recently, an ontology-driven decision tree learning algorithm was proposed [24] to learn classification rules at multiple levels of abstraction. Although called ontology-driven, what the proposed solution really uses is a taxonomy, i.e., a set of IS-A relations associated with each attribute. It consists in a top-down concept hierarchy guided search in a hypothesis space of decision trees.

A variation of the Naïve Bayes Learner making use of attribute-value taxonomies (AVT-NBL) was proposed [23]. It starts with the Naïve Bayes classifier based on the most abstract value for each attribute, and successively moves in the direction of the more concrete values, i.e., the ones appearing in the data set. The idea is to stop somewhere in between, in order to achieve a balance between the complexity of the resulting classifier and its classification accuracy.

It suffers from some of the same problems of EG2, as the authors never specify a standard format to represent the domain knowledge and the knowledge that can be represented is restricted to IS-A relations, a small subset of an ontology. It is however a tentative step in a meaningful direction, facilitating the discovery of classifiers at different levels of abstraction.

3   Preliminaries

As far as the authors know existing approaches to introducing some form of domain knowledge in the classification process deal only with taxonomies which are a fraction of the expressive power of true ontologies.

We need a standard way of expressing domain knowledge, so it can be shared and reused. The Web Ontology Language, OWL 2 [13] satisfies this criterion and offers plenty expressive power to use in the context of classification. We assume that the reader is somewhat familiar with OWL 2 and with the Manchester syntax. Nonetheless we briefly review the main components of an OWL 2 ontology.

3.1   OWL 2

The main components of an OWL 2 ontology are axioms, classes, individuals and properties. Two types of properties exist: data properties have a literal as a range and object properties have a class has range.

Note that classes in the ontology have nothing to do with class of the instance in a classification problem, i.e., the attribute value we are trying to predict.

Classes provide an abstraction mechanism for grouping resources with similar characteristics. When you think of the concept *Parasol*, for instance, you are not thinking of any concrete mushroom. Rather all the mushrooms that share the necessary characteristics to be considered of that species. However if you embark in a mushroom hunting adventure you will probably find a mushroom of this species for dinner. That mushroom is an Individual of the class *Parasol*.

Characteristics are called properties and odor is an example of a property of the class *Parasol*.

So how can one define which individuals belong to the class *Parasol*? Using axioms. Axioms are the core of an OWL 2 ontology and are essentially statements that are truth in the domain. You can then say that mushrooms with white spore print color and not white gills are of the class *Parasol*. You can also say that *Parasol* is a subclass of *Mushroom*, i.e., all individuals in the class *Parasol* are also in the class *Mushroom*.

However reasoning in a OWL DL ontology is a problem in NEXPTIME which is highly undesirable for our intended application.

3.2   OWL 2 EL

Fortunately a subset exists that trades off some aspects of the OWL DL expressive power in return for PTIME complexity in all the standard reasoning tasks. From now on, whenever we refer to OWL or to ontologies, keep in mind that we are talking about OWL 2 EL.

We wish to have a set of axioms that define class membership and then quickly compute which individuals belong to which classes. Although it might sound simple, it is a rather complex topic and an area of research in itself.

3.3   ELK

We use Elk [10] to determine which individuals in the ontology belong to which classes. This is called ABox Realization.

Realization is the task of computing the implied instance/type relationships between all named individuals and named classes in an ontology. Only direct instance/type relations are returned in the result. In order to determine which instance/type relations are direct, one needs to know all subclass/superclass between named classes in the ontology. Therefore, ELK automatically triggers TBox classification before ABox realization.

TBox classification is the task of computing the implied subclass/super-class relationships between all named classes in an ontology. Besides finding out whether a class is subsumed by another one or not, this task involves the transitive reduction of the computed class taxonomy: only direct subclass/superclass relations are returned in the result.

4   Structuring an Ontology to Support Classification Problems

Until now we have presented ontologies as a completely separated topic from the problem of learning to predict a target attribute from a set of labeled examples.

In this section we will make a bridge between the data set and the ontology, so the algorithm can leverage the available domain knowledge and the labeled instances in the data set to produce a more precise and compact model.

Attribute values in the data set that we want to use while defining axioms are added as Individuals to the ontology. Consider that we are interested in the following odors: creosote (c), fishy (y), foul (f), musty (m), pungent (p) and spicy (s) which are bad smells but not in almond (a), anise (l) or none (n). Also consider that we are only interested in green (r) spore-print-color.

OWL fragment 1.1. What mushrooms odors smell bad?

```
1   Class: BadSmell
2
3   Individual: c
4       Types: BadSmell
5   Individual: y
6       Types: BadSmell
7   Individual: f
8       Types: BadSmell
9   Individual: m
10      Types: BadSmell
11  Individual: p
12      Types: BadSmell
13  Individual: s
14      Types: BadSmell
```

OWL fragment 1.2. Green spore print colors make greenish mushrooms

```
1   Class: Greenish
2
3   Individual: r
4       Types: Greenish
```

Existing attributes (in the data set) that we wish to mention in our axioms are added as object properties. Suppose that we are interested in odor and in spore-print-color.

OWL fragment 1.3. Definition of the attributes *odor* and *spore print color* in the ontology, allowing the definition of axioms that use these attributes

```
1  ObjectProperty:  odor
2  ObjectProperty:  spore−print−color
```

A *meta*-class "Attribute" that can have two kinds of direct subclasses. New attributes have no corresponding object property and represent a new dimension in which instances in the data set can be considered. These kind of new attributes result from the application of a set of axioms to the existing attribute values or to an abstraction of them. In the next example we will add a new attribute called *Species*.

On the other hand, direct subclasses of the *meta*-class "Attribute" that have a corresponding object property represent attributes that already exist in the data set but will have multiple levels of abstraction. Each direct subclass of one of this attributes represent a new level of abstraction to be considered. In the next example we will add a higher level of abstraction to the attribute odor, called smell. The subclasses of smell are the possible attribute values of the new attribute smell.

OWL fragment 1.4. Attribute hierarchy showing a new class *species* and an higher level of abstraction *smell* for the attribute *odor*

```
 1  Class:  Odor
 2  Class:  Smell
 3      SubClassOf:  Odor
 4  Class:  Species
 5  Class:  Attribute
 6      SuperClassOf:  Odor
 7      SuperClassOf:  Species
 8
 9  Class:  BadSmell
10      SubClassOf:  Smell
```

Note that the subclasses of smell and species are the possible attribute values. We can have how many attributes values we want. However note that it is possible that some instances are not part if any of these attribute values because they are not part of any of the corresponding classes. As an example, consider any instance where the attribute value of the attribute odor is anise. This instance is not part of the BadSmell class and there are no other subclasses of odor. When this happens the attribute in question will have a new special attribute value "NA" that will have all instances that do not belong to any attribute value. One might be tempted to define a new class GoodSmell as the negation of the class BadSmell. This is a violation of the EL profile as it does not support class negation.

At last, suppose that we know that if some mushroom smells bad or has greenish spore print color it is of the species "FalseParasol".

OWL fragment 1.5. What characteristics must a mushroom have in order to belong to the species *False parasol?*

1	**Class**: FalseParasol
2	**SubClassOf**: Species
3	**SuperClassOf**: odor **some** BadSmell
4	**SuperClassOf**: spore−print−color **some** Greenish

5   Ontology Aware Decision Tree Learner

Now that we have a bridge between instances in the data set and domain knowledge in the ontology we will enrich each instance in the data set with what we can infer from the ontology. Suppose that we have an instance with green spore print color and a poignant odor. From the ontology we know that the species of this instance is "FalseParasol" and that it smells bad.

Algorithm 1 creates an individual in the ontology for each instance in the data set and makes object property assertions corresponding to the instance attribute values. After it is run all instances are classified. Note that instances in the data set are projected into the ontology as individuals and only the attribute values that can influence class inference are added. In our example the individuals added to the ontology would have only two dimensions: odor and spore print color.

Also note that by leveraging incremental reasoning the inner loop does not trigger a full re-computation. This step can be completed in PTIME.

In Algorithm 2, for each new attribute (as defined in the ontology), we fetch the individuals for each possible attribute value. After this step we can proceed to attribute selection as we usually would in a normal decision tree algorithm. In our implementation we use a simple version of the ID3 algorithm.

Algorithm 1. Projects data set instances into the ontology as individuals

---

```
 1 procedure PROJECTTOONTOLOGY(instances, ontology)
 2    for all i ∈ instances do
 3       j ← individual(i)                    ▷ create individual for instance
 4       ontology ← ontology + j              ▷ add individual to ontology
 5       for all a ∈ attributes(i) do
 6          v ← value(i, a)
 7          if hasProperty(a, ontology) ∧ hasIndividual(v, ontology) then
 8             objectPropertyAssertion(j, a, v, ontology)
 9          end if
10       end for
11    end for
12 end procedure
```

Algorithm 2. Obtains attribute values for the new generated attributes

---

```
 1 procedure GETATTRIBUTEVALUES(ontology)
 2     for all a ∈ subClassOf('Attribute', ontology, direct = True) do
 3         if hasProperty(a, ontology) then       ▷ higher levels of abstraction for a
 4             for all aₕ ∈ subClassOf(a, ontology, direct = True) do
 5                 for all v ∈ subClassOf(aₕ, ontology, direct = True) do
 6                     instances(aₕ, v) ← individuals(v, ontology)
 7                 end for
 8             end for
 9         else                                   ▷ a is not an abstraction of an existing attribute
10             for all v ∈ subClassOf(a, ontology, direct = True) do
11                 instances(a, v) ← individuals(v, ontology)
12             end for
13         end if
14     end for
15 end procedure
```

5.1 Attribute Selection Criterion

The information gain $IG(A_i)$ of an attribute A_i is calculated as follows:

$$IG(A_i) = H(T) - \sum_{f \in F(A_i)} p(f) H(t_{A_i = f}) \tag{1}$$

where $H(T)$ is the entropy of the training set and $H(t_{A_i=f})$ is the entropy of a subset of the training set formed by the instances of T where the value of attribute A_i is f. The entropy of a set T is given by:

$$H(T) = - \sum_{c_j \in C} p(c_j) \log_2 p(c_j) \tag{2}$$

One might think that as ID3 recursively selects the attribute with the highest information gain it would naturally pick attributes at higher levels of abstraction if those led to simpler, more accurate trees.

This is, however, not the case. The information gain metric is biased through attributes with a greater number of features [22].

Proposition 1. *Given an attribute A_i and an attribute A_{1i} where at least one feature of A_{1i} is at an higher level of abstraction and all others are at least at the same level then $IG(A_i) \geq IG(A_{1i})$.*

Proof. The case where exactly one feature from A_i appears in A_{1i} at an higher level of abstraction and all other remain the same is a mere renaming of one feature in practical terms and is trivial to observe that no counts change because of it and consequently, in this case, $IG(A_i) = IG(A_{1i})$.

Now consider the case where n features $f_1, \ldots, f_n$ from A_i are represented by a common ancestor f_a in A_{1i} and all other features remain the same. Equivalently we might say that A_i can be obtained from A_{1i} by splitting f_a in n features. This is exactly the case where it has been shown [16,22] that the information gain of the attribute with more features is greater than or equal to the attribute with less features even if the features of the later are already sufficiently fine for the induction task at hand.

This is highly undesirable when dealing with attributes at different levels of abstraction. As we climb up in the feature hierarchy, more features will be aggregated and the attribute representing that level of abstraction will consequently have less possible values. Using information gain this attribute will never be selected.

The gain ratio attribute selection measure [16] minimizes this bias and can be calculated as follows:

$$\frac{IG(A_i)}{-\sum_{f \in F(A_i)} p(f) \log_2 p(f)} \tag{3}$$

5.2   Ontology Aware Decision Tree Classifier

The model produced can be used to classify instances where most of the attributes are missing as long as there is enough information to infer the value of the new attributes and together with the existing ones they are enough to reach a leaf of the decision tree.

Given an instance I with some possibly missing values we compute new attribute values using a method analogous to the one described in the previous section and obtain an extended version of I. We use this extended version to navigate the decision tree as usually.

6   Results

In order to execute some experiments and compare the performance of the proposed algorithm with standard ID3 decision tree algorithm a Java implementation was developed, as part of the D2PM framework [2].

In spite of data with values specified at different levels of abstraction being common in many domains of application there are few standard benchmark data sets with these characteristics and with an associated ontology. We selected the *Mushroom* data set from the UCI Machine Learning Repository as a starting point.

Domain knowledge obtained from the book "The Mushroom Hunter's Field Guide" was made explicit in an OWL 2 ontology.

Three sets of experiments were then executed. The first compares the accuracy of the proposed Ontology Aware Decision Tree with the standard ID3 algorithm on the original data, where all values are concrete. We also look at the complexity of the produced decision trees. Figure 1 shows an example of a decision tree generated by OADT for randomly selected small training sets. This simple tree has an accuracy over the entire data set of **0.914** while the standard

Table 1. Accuracy and tree size of ID3 and OADT on the original data

	ID3 accuracy	OADT accuracy	ID3 tree size	OADT tree size
S1	0.785	0.999	5	4
S2	0.999	1.000	7	4
S3	0.999	1.000	7	4
S4	0.993	0.999	5	4
S5	1.000	1.000	7	4
Avg.	0.9552	0.9996	6.2	4

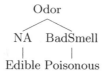

Fig. 1. Example of a decision tree generated by OADT from a small training set (<50 instances)

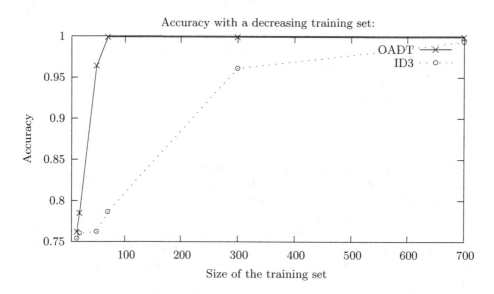

Fig. 2. The influence of training set size on the accuracy of ID3 and OADT.

ID3 algorithm, for the same training set, generates a tree that has an accuracy of only **0.549**.

The second shows how the accuracy of both algorithms changes with increasingly smaller training sets. A subset with 1000 instances was randomly selected

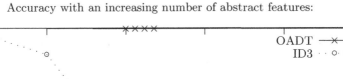

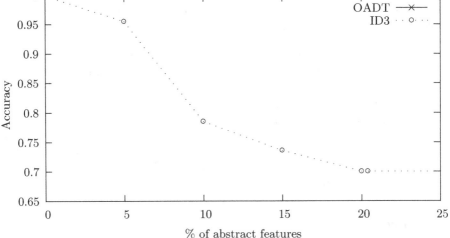

Fig. 3. Accuracy of ID3 and OADT in a data set with an increasing number of abstract values.

from the original data set to serve as a test set. Six subsets were randomly selected from the remaining instances of the original data set, with sizes of 700, 300, 70, 50, 20 and 15 instances to be used as training sets. The results are shown in Fig. 2.

The third studies how the accuracy of both algorithms changes with an increasing number of values being abstract, e.g., not knowing the exact *odor* of a mushroom but being able to tell if it has a pleasant or bad smell. Starting from a data set with no abstract values, five data sets were then generated with an approximate percentage of abstract values of 5 %, 10 %, 15 %, 20 % and 25 %. Figure 3 shows these results.

To assess the accuracy of the two algorithms on the original data set, we used cross-validation by repeated random sub-sampling. Five disjoint subsets were randomly selected and each was divided in two disjoint subsets, a training set and a test set. Table 1 shows these results.

These results are in line with our expectations. First, even on data were all values are concrete, domain knowledge can help build models that perform as good or better while being considerably simpler. This difference in accuracy is more pronounced with smaller training sets.

Second, when the specific concrete values are unknown but a more abstract version is available, OADT maintains its performance remarkably well while the performance of traditional ID3 decreases as more values are expressed at higher levels of abstraction.

7 Summary

In this paper, we have described an approach for learning from ontologies, represented in a standard knowledge representation language, OWL 2 EL, and from a set of training data.

We show that it is possible to produce simpler decision trees that perform as good or better than traditional approaches. We also show that domain knowledge helps produce classifiers that perform very well in small training sets comparatively to traditional algorithms.

We also describe how to change the classifier to leverage available knowledge to classify instances with values not seen in the training set, by inferring values from the ontology. This is useful to classify instances where the concrete value of some attributes is not known but the abstract class is, e.g., not knowing the exact phylum of a bacteria, but knowing it is Gram positive.

References

1. Altman, N.: An introduction to kernel and nearest-neighbor nonparametric regression. Am. Stat. **46**(3), 175–185 (1992)
2. Antunes, C.: D2PM: domain driven pattern mining. Technical report, Project report, Technical Report 1530, IST, Lisboa (2011)
3. Bache, K., Lichman, M.: UCI machine learning repository (2013). http://archive.ics.uci.edu/ml
4. Blockeel, H., De Raedt, L.: Top-down induction of first-order logical decision trees. Artif. Intell. **101**(1), 285–297 (1998)
5. Boser, B.E., Guyon, I.M., Vapnik, V.N.: A training algorithm for optimal margin classifiers. In: Proceedings of the Fifth Annual Workshop on Computational Learning Theory, pp. 144–152. ACM (1992)
6. Bramer, M.: Using J-pruning to reduce overfitting in classification trees. Knowl.-Based Syst. **15**(5), 301–308 (2002)
7. Domingos, P., Kok, S., Poon, H., Richardson, M., Singla, P.: Unifying logical and statistical AI. In: AAAI (2006)
8. Dzeroski, S., Jacobs, N., Molina, M., Moure, C., Muggleton, S., Laer, W.V.: Detecting traffic problems with ILP. In: Page, D.L. (ed.) ILP 1998. LNCS, vol. 1446, pp. 281–290. Springer, Heidelberg (1998)
9. Hawkins, D.M.: The problem of overfitting. J. Chem. Inf. Comput. Sci. **44**(1), 1–12 (2004)
10. Kazakov, Y., Krtzsch, M., Simančík, F.: The incredible ELK. J. Autom. Reasoning **53**(1), 1–61 (2014). http://dx.doi.org/10.1007/s10817-013-9296-3
11. Lincoff, G., Nehring, C.: National Audubon Society Field Guide to North American Mushrooms. Knopf, New York (1997)
12. Maimon, O., Rokach, L. (eds.): Data Mining and Knowledge Discovery Handbook, 2nd edn. Springer, New York (2010)
13. Motik, B., Patel-Schneider, P.F., Parsia, B., Bock, C., Fokoue, A., Haase, P., Hoekstra, R., Horrocks, I., Ruttenberg, A., Sattler, U., et al.: Owl 2 web ontology language: structural specification and functional-style syntax. W3C recommendation 27, 17 (2009)

14. Muggleton, S., De Raedt, L., Poole, D., Bratko, I., Flach, P., Inoue, K., Srinivasan, A.: ILP turns 20. Mach. Learn. **86**(1), 3–23 (2012)
15. Núñez, M.: The use of background knowledge in decision tree induction. Mach. Learn. **6**(3), 231–250 (1991)
16. Quinlan, J.R.: Induction of decision trees. Mach. Learn. **1**(1), 81–106 (1986)
17. Quinlan, J.R., Cameron-Jones, R.M.: Foil: a midterm report. In: Brazdil, P.B. (ed.) ECML 1993. LNCS, vol. 667, pp. 1–20. Springer, Heidelberg (1993)
18. Quinlan, J.R.: C4.5: Programs for Machine Learning, vol. 1. Morgan Kaufmann, San Francisco (1993)
19. Roberts, S., Jacobs, N., Muggleton, S., Broughton, J., et al.: A comparison of ILP and propositional systems on propositional traffic data. In: Page, D.L. (ed.) ILP 1998. LNCS, vol. 1446, pp. 291–299. Springer, Heidelberg (1998)
20. Rumelhart, D.E., Hinton, G.E., Williams, R.J.: Learning representations by back-propagating errors. Cogn. Model. **1**, 213 (2002)
21. Srinivasan, A., King, R.D., Muggleton, S.: The role of background knowledge: using a problem from chemistry to examine the performance of an ILP program. Trans. Knowl. Data Eng. (1999)
22. White, A.P., Liu, W.Z.: Technical note: Bias in information-based measures in decision tree induction. Mach. Learn. **15**(3), 321–329 (1994)
23. Zhang, J., Kang, D.K., Silvescu, A., Honavar, V.: Learning accurate and concise naïve bayes classifiers from attribute value taxonomies and data. Knowl. Inf. Syst. **9**(2), 157–179 (2006)
24. Zhang, J., Silvescu, A., Honavar, V.G.: Ontology-driven induction of decision trees at multiple levels of abstraction. In: Koenig, S., Holte, R. (eds.) SARA 2002. LNCS (LNAI), vol. 2371, pp. 316–323. Springer, Heidelberg (2002)

A System for Tractable Computation of Negative Answers to Conjunctive Queries

Jianfeng Du$^{(\boxtimes)}$, Deqian Liu, Can Lin, Xuzhi Zhou, and Jiaqi Liang

Guangdong University of Foreign Studies, Guangzhou 210096, China
jfdu@gdufs.edu.cn

Abstract. Users of conjunctive query answering (CQA) in description logic (DL) ontologies may often confront a problem that there are insufficiently many or even no certain answers to a given conjunctive query (CQ). It is needed to suggest users negative answers to a given CQ that are expected to be in the result but actually do not occur there while explaining why these answers are not certain ones. To this end, we develop a system which incrementally computes negative answers in a user interactive mode for a given CQ and a given consistent DL ontology. A negative answer that is incrementally computed in level k is a tuple such that there is a set of k assertions whose appending to the ontology makes this tuple a certain answer while keeping the ontology consistent. The proposed computational method works in polynomial time in terms of data complexity for DL ontologies with first-order rewritable TBoxes.

1 Introduction

The W3C proposal of the Web Ontology Language (OWL) has encouraged more and more ontology-based applications. One typical application is ontology-based data access (OBDA), which enables new information systems that use an ontology as a unified view of the information in variant data sources. Conjunctive query answering (CQA) is a common service provided by information systems such as OBDA systems for accessing explicit and implicit information. Users of CQA may often confront a problem that there are insufficiently many or even no certain answers to a given CQ. In this situation, users may want to see why some tuples (called *negative answers*) that are expected to be in the result are not certain answers to the CQ. This requirement has been raised by [2] for improving the usability of OBDA systems. In other words, it is needed to suggest users negative answers to a given CQ while explaining why these answers are not certain ones.

We target a common scenario where the given ontology is expressed in description logics (DLs) [1] that underpin OWL while a negative answer is explained by a set of assertions which turns the negative answer into a certain answer and keeps the ontology consistent. Although there exists work on analyzing the complexity for explaining a negative answer [5] and work on computing representative explanations for a negative answer [7], the problem of computing negative answers and their corresponding explanations is rarely investigated.

D. Zhao et al. (Eds.): CSWS 2014, CCIS 480, pp. 56–66, 2014.
DOI: 10.1007/978-3-662-45495-4_5

A preliminary attempt has been made in [8] for computing negative answers and some of their explanations. The approach proposed in [8] relies on backward inference and search of hypothetical assertions. It only works for DL ontologies that can be directly translated to function-free Horn rules, and has a high computational complexity beyond polynomial time in terms of *data complexity*, namely the complexity measured in the number of assertions only.

In this paper we aim at a different approach to computing negative answers and their corresponding explanations. It works for a common class of DLs covering most languages in the DL-Lite family [4] and is *tractable* (i.e. works in polynomial time) in terms of data complexity. The main contributions are as follows. Firstly, we propose a new class of explanations, called $\subset_{ms}$-*explanations*, which is more general than the class of representative explanations proposed in [7] and enables a more efficient computational method. Secondly, we propose a new problem for query answering, called the *more-answers* problem, which is more general than the CQA problem and can be used to incrementally compute certain and negative answers. Thirdly, we propose a tractable method (in data complexity) for solving the more-answers problem. Finally, we develop a system for incrementally computing answers in a user interactive mode. All answers that have explanations (including certain answers) are incrementally computed by solving a sequence of instances of the more-answers problem. Experimental results show that the system scales up to millions of assertions.

The remainder of this paper is organized as follows. Section 2 provides background knowledge about description logics (DLs), conjunctive query and first-order rewritable TBox. Section 3 presents the semantics and a tractable method for computing all $\subset_{ms}$-explanations. Section 4 gives the semantics and a tractable method for solving the more-answers problem. Before concluding in Sect. 7, Sect. 6 describes our experimental evaluation.

2   Preliminaries

We assume that the reader is familiar with DLs [1]. A DL ontology consists of a TBox and an ABox. We assume that the *Unique Name Assumption* [1] is adopted and only consider ABoxes consisting of *basic assertions*, namely concept assertions $A(a)$ and role assertions $r(a, b)$, where A is a concept name, r is a role name, and a and b are individuals. Other concept assertions and role assertions can be normalized to basic ones in a standard way. Let Σ be a set of concept names and role names. An ABox that contains only concept names and role names from Σ is called a Σ-ABox. We use the traditional semantics for DLs given in [1]. A DL ontology $\mathcal{O}$ is said to be *consistent*, denoted by $\mathcal{O} \not\models \bot$, if it has at least one model; otherwise *inconsistent*, denoted by $\mathcal{O} \models \bot$.

A *conjunctive query* (CQ) is of the form $Q(\overrightarrow{x}) = \exists \overrightarrow{y} : \phi(\overrightarrow{x}, \overrightarrow{y}, \overrightarrow{c})$, where $\phi(\overrightarrow{x}, \overrightarrow{y}, \overrightarrow{c})$ is a conjunction of atoms over concept names, role names and equality, where $\overrightarrow{x}$ are *answer variables*, $\overrightarrow{y}$ *quantified variables* and $\overrightarrow{c}$ individuals. A CQ without answer variables is a called a *Boolean conjunctive query* (BCQ). A CQ can also be treated as a set of atoms. By $|S|$ we denote the cardinality of a set S. A *substitution* for a set S of atoms is a mapping from variables in S to individuals or variables; it is called *ground* (resp. *variant*) if it maps variables in S to individuals (resp. variables) only. A tuple $\overrightarrow{t}$ of individuals is called a *certain answer* to a CQ $Q(\overrightarrow{x})$ in an ontology $\mathcal{O}$ if the arity of $\overrightarrow{t}$ agrees with that of $\overrightarrow{x}$ and the BCQ $Q(\overrightarrow{t})$ is satisfied by all models of $\mathcal{O}$, denoted by $\mathcal{O} \models Q(\overrightarrow{t})$. Otherwise if $\mathcal{O} \not\models Q(\overrightarrow{t})$, $\overrightarrow{t}$ is called a *negative answer* to $Q(\overrightarrow{x})$ in $\mathcal{O}$. The set of certain answers to $Q(\overrightarrow{x})$ in $\mathcal{O}$ is denoted by $\mathsf{ans}(\mathcal{O}, Q(\overrightarrow{x}))$.

Some DLs can be translated to Datalog$^\pm$ [3]. A Datalog$^\pm$ ontology consists of finitely many *tuple generating dependencies* (TGDs) $\forall \overrightarrow{x} \forall \overrightarrow{y} : \phi(\overrightarrow{x}, \overrightarrow{y}) \rightarrow \exists \overrightarrow{z} : \varphi(\overrightarrow{x}, \overrightarrow{z})$, *constraints* $\forall \overrightarrow{x} : \phi'(\overrightarrow{x}) \rightarrow \bot$, as well as *equality generating dependencies* (EGDs) $\forall \overrightarrow{x} : \phi'(\overrightarrow{x}) \rightarrow x_1 = x_2$, where $\phi(\overrightarrow{x}, \overrightarrow{y})$, $\varphi(\overrightarrow{x}, \overrightarrow{z})$ and $\phi'(\overrightarrow{x})$ are conjunctions of atoms, x_1 and x_2 occur in $\overrightarrow{x}$, and $\bot$ denotes the truth constant false. The portions of a TBox $\mathcal{T}$ that are translated to TGDs, constraints and EGDs are denoted by $\mathcal{T}_D$, $\mathcal{T}_C$ and $\mathcal{T}_E$, respectively. A TBox $\mathcal{T}$ is said to be *first-order rewritable* if it can be translated to a Datalog$^\pm$ ontology with the following conditions satisfied for any CQ $Q(\overrightarrow{x})$ and any $\Sigma_{\mathcal{T}}$-ABox $\mathcal{A}$, where $\Sigma_{\mathcal{T}}$ is the set of concept names and role names in $\mathcal{T}$:

(1) $\mathsf{ans}(\mathcal{T} \cup \mathcal{A}, Q(\overrightarrow{x})) = \mathsf{ans}(\mathcal{T}_D \cup \mathcal{A}, Q(\overrightarrow{x}))$ when $\mathcal{T} \cup \mathcal{A} \not\models \bot$;
(2) $\mathcal{T}_C \cup \mathcal{T}_E$ can be rewritten to a finite set of BCQs according to $\mathcal{T}_D$, denoted by $\gamma(\mathcal{T}_C \cup \mathcal{T}_E, \mathcal{T}_D)$, such that $\mathcal{T} \cup \mathcal{A} \models \bot$ if and only if $\mathcal{A} \models Q'$ for some $Q' \in \gamma(\mathcal{T}_C \cup \mathcal{T}_E, \mathcal{T}_D)$;
(3) $Q(\overrightarrow{x})$ can be rewritten to a finite set of CQs of the form $Q'(\overrightarrow{x})$ according to $\mathcal{T}_D$, denoted by $\tau(Q(\overrightarrow{x}), \mathcal{T}_D)$, such that $\mathsf{ans}(\mathcal{T}_D \cup \mathcal{A}, Q(\overrightarrow{x})) = \bigcup_{Q'(\overrightarrow{x}) \in \tau(Q(\overrightarrow{x}), \mathcal{T}_D)} \mathsf{ans}(\mathcal{A}, Q'(\overrightarrow{x}))$.

The TBoxes expressed by most DLs in the DL-Lite family are first-order rewritable [3]. Since the DL-Lite family has become popular in information systems such as OBDA systems, restricting TBoxes to be first-order rewritable does not matter in many applications. We only consider first-order rewritable TBoxes.

In the following a running example about the aforementioned notions is given.

Example 1. Let $\mathcal{O} = \mathcal{T} \cup \mathcal{A}$ be a consistent ontology where the TBox $\mathcal{T}$ consists of the following axioms $\alpha_1, \ldots, \alpha_5$ and the ABox $\mathcal{A}$ consists of the following axioms α_6 and α_7.

α_1 : Employee $\sqsubseteq$ Person (*An employee is a person.*)
α_2 : Employee $\sqsubseteq$ worksFor.$\top$ (*An employee works for something.*)
α_3 : worksFor.$\top \sqsubseteq$ Employee (*Something that works is an employee.*)
α_4 : worksFor$^-$.$\top \sqsubseteq$ Company (*What something works for is a company.*)
α_5 : Employee $\sqsubseteq \neg$Company (*Employees are not companies.*)
α_6 : Person(Tom) (*Tom is a person.*)
α_7 : worksFor(Amy, IBM) (*Amy works for IBM.*)

Given a CQ $Q(\langle x \rangle) = \exists y : \mathsf{Person}(x) \wedge \mathsf{worksFor}(x, y)$, which finds all persons that work for something, we can check that $\langle \mathsf{Amy} \rangle$ is the only certain answer to $Q(\langle x \rangle)$ in $\mathcal{O}$, while $\langle \mathsf{Tom} \rangle$ and $\langle \mathsf{IBM} \rangle$ are two negative answers to $Q(\langle x \rangle)$ in $\mathcal{O}$.

The TBox $\mathcal{T}$ is actually first-order rewritable, where $\mathcal{T}_D = \{\alpha_1, \alpha_2, \alpha_3, \alpha_4\}$, $\mathcal{T}_C = \{\alpha_5\}$ and $\mathcal{T}_E = \emptyset$. The rewritten results for $\mathcal{T}_C \cup \mathcal{T}_E$ according to $\mathcal{T}_D$ can be $\gamma(\mathcal{T}_C \cup \mathcal{T}_E, \mathcal{T}_D) = \{Q'_1, Q'_2, Q'_3, Q'_4\}$. The rewritten results for $Q(\langle x \rangle)$ according to $\mathcal{T}_D$ can be $\tau(Q(\langle x \rangle), \mathcal{T}_D) = \{Q_1(\langle x \rangle), Q_2(\langle x \rangle)\}$.

- $Q'_1 = \exists x \colon \mathsf{Employee}(x) \wedge \mathsf{Company}(x)$
- $Q'_2 = \exists x, y \colon \mathsf{Employee}(x) \wedge \mathsf{worksFor}(y, x)$
- $Q'_3 = \exists x, y \colon \mathsf{worksFor}(x, y) \wedge \mathsf{Company}(x)$
- $Q'_4 = \exists x, y, z \colon \mathsf{worksFor}(x, y) \wedge \mathsf{worksFor}(z, x)$
- $Q_1(\langle x \rangle) = \mathsf{Employee}(x)$
- $Q_2(\langle x \rangle) = \exists y \colon \mathsf{worksFor}(x, y)$

3   The $\sqsubset_{\mathsf{ms}}$-Explanations

In [7] the query abduction problem for BCQs is studied. Given a consistent DL ontology $\mathcal{O} = \mathcal{T} \cup \mathcal{A}$, a BCQ Q and a set Σ of concept names and role names (called *abducible predicates*), $\mathcal{P} = (\mathcal{T}, \mathcal{A}, Q, \Sigma)$ is called an instance of the *query abduction problem*. An *explanation* for $\mathcal{P}$ is a Σ-ABox $\mathcal{E}$ such that $\mathcal{T} \cup \mathcal{A} \cup \mathcal{E} \models Q$ and $\mathcal{T} \cup \mathcal{A} \cup \mathcal{E} \not\models \bot$. Since an explanation may contain fresh individuals not in $\mathcal{P}$ (i.e. neither in $\mathcal{O}$ nor in Q), a variant set-inclusion minimality is defined by treating fresh individuals as variables. A *substitution* for an explanation $\mathcal{E}$ is a mapping from fresh individuals in $\mathcal{E}$ to existing or fresh individuals. A *renaming* for $\mathcal{E}$ is a substitution for $\mathcal{E}$ that maps different fresh individuals to different fresh individuals. For two explanations $\mathcal{E}$ and $\mathcal{E}'$, $\mathcal{E}' \sqsubset_{\mathsf{r}} \mathcal{E}$ denotes that there exists a renaming ρ for $\mathcal{E}'$ such that $\mathcal{E}'\rho \subset \mathcal{E}$. A *minimal explanation* $\mathcal{E}$ for $\mathcal{P}$ is an explanation for $\mathcal{P}$ such that there is no explanation $\mathcal{E}'$ for $\mathcal{P}$ fulfilling $\mathcal{E}' \sqsubset_{\mathsf{r}} \mathcal{E}$.

Here we introduce a new class of explanations, defined below.

Definition 1 ($\sqsubset_{\mathsf{ms}}$-explanation). *Let $\sqsubset_{\mathsf{ms}}$ be a precedence relation between explanations for $\mathcal{P}$ such that $\mathcal{E}' \sqsubset_{\mathsf{ms}} \mathcal{E}$ if and only if $\mathcal{E}'\sigma \subseteq_{\mathsf{m}} \mathcal{E}$ for some substitution σ while $\mathcal{E}\theta \not\subseteq_{\mathsf{m}} \mathcal{E}'$ for any substitution θ, where $S' \subseteq_{\mathsf{m}} S$ denotes that S' is a subset of S when both sets S' and S are treated as multisets, i.e. sets in which the elements may appear more than once. An explanation $\mathcal{E}$ for $\mathcal{P}$ is called a $\sqsubset_{\mathsf{ms}}$-explanation for $\mathcal{P}$ if there is no explanation $\mathcal{E}'$ for $\mathcal{P}$ such that $\mathcal{E}' \sqsubset_{\mathsf{ms}} \mathcal{E}$. A $\sqsubset_{\mathsf{ms}}$-explanation $\mathcal{E}$ for $\mathcal{P}$ with $|\mathcal{E}| = k$ is also called a $\sqsubset^k_{\mathsf{ms}}$-explanation for $\mathcal{P}$. The set of different $\sqsubset^k_{\mathsf{ms}}$-explanations for $\mathcal{P}$ up to renaming of fresh individuals (simply renaming) is denoted by $\sqsubset^k_{\mathsf{ms}}$-expl$(\mathcal{P})$.*

Since $\mathcal{E}' \sqsubset_{\mathsf{r}} \mathcal{E}$ implies $\mathcal{E}' \sqsubset_{\mathsf{ms}} \mathcal{E}$, $\sqsubset_{\mathsf{ms}}$-explanations are minimal explanations. A $\sqsubset_{\mathsf{ms}}$-explanation is similar to a *representative explanation* which is defined in [7] as a minimal explanation $\mathcal{E}$ such that $\mathcal{E}' \sqsubset_{\mathsf{s}} \mathcal{E}$ for no minimal explanation $\mathcal{E}'$, where $\sqsubset_{\mathsf{s}}$ is a precedence relation between explanations such that $\mathcal{E}' \sqsubset_{\mathsf{s}} \mathcal{E}$ if and only if $\mathcal{E}'\sigma \subseteq \mathcal{E}$ for some substitution σ while $\mathcal{E}\theta \not\subseteq \mathcal{E}'$ for any substitution θ. The definition of $\sqsubset_{\mathsf{ms}}$-explanation differs from that of representative explanation in using the multiset-based subset relation $\subseteq_{\mathsf{m}}$ other than the traditional subset relation $\subseteq$. In fact, representative explanations are $\sqsubset_{\mathsf{ms}}$-explanations because

$\mathcal{E}' \sqsubseteq_{ms} \mathcal{E}$ implies $\mathcal{E}' \sqsubseteq_s \mathcal{E}$ when $\mathcal{E}$ is a minimal explanation. In other words, the class of minimal explanations is more general than the class of $\sqsubseteq_{ms}$-explanations, while the class of $\sqsubseteq_{ms}$-explanations is more general than the class of representative explanations. This relation is formalized in the following proposition.

Proposition 1. *A representative explanation for $\mathcal{P}$ is also a $\sqsubseteq_{ms}$-explanation for $\mathcal{P}$. A $\sqsubseteq_{ms}$-explanation for $\mathcal{P}$ is also a minimal explanation for $\mathcal{P}$.*

The precedence relation $\sqsubseteq_{ms}$ enables an efficient method for computing $\sqsubseteq_{ms}^k$-expl($\mathcal{P}$) without considering $\sqsubseteq_{ms}$-explanations with larger cardinalities and without computing minimal explanations beforehand. Similarly as [7], we define a *bipartition* of a CQ $Q(\overrightarrow{x})$ as a tuple of two CQs $\langle Q_1(\overrightarrow{x_1}), Q_2(\overrightarrow{x_2}) \rangle$ such that $Q_1(\overrightarrow{x_1}) \cap Q_2(\overrightarrow{x_2}) = \emptyset$ and $Q_1(\overrightarrow{x_1}) \cup Q_2(\overrightarrow{x_2}) = Q(\overrightarrow{x})$. For $\mathcal{P} = (\mathcal{T}, \mathcal{A}, Q, \Sigma)$, let $\Xi_k(\mathcal{P}) = \{Q_1\theta \mid Q' \in \tau(Q, \mathcal{T}_D), \langle Q_1, Q_2 \rangle$ is a bipartition of Q' such that $|Q_1| = k$ and Q_1 contains only predicates in Σ, and θ is a ground substitution for Q_2 such that $Q_2\theta \subseteq \mathcal{A}\}$, and $\Gamma_k(\mathcal{P}) = \{E\sigma \mid E \in \Xi_k(\mathcal{P})$ and σ is a fresh substitution for E in $\mathcal{P}$ such that $\mathcal{T} \cup \mathcal{A} \cup E\sigma \not\models \bot\}$, where a *fresh substitution* for E in $\mathcal{P}$ is a ground substitution for E that only maps variables in E to fresh individuals not in $\mathcal{P}$. The following theorem shows a sound and complete method for computing $\sqsubseteq_{ms}^k$-expl($\mathcal{P}$) using the above notations, where $S' \doteq S$ denotes that two sets S' and S coincide up to renaming.

Theorem 1. $\sqsubseteq_{ms}^k$-expl($\mathcal{P}$) $\doteq \{\mathcal{E} \in \Gamma_k(\mathcal{P}) \mid$ *there is no non-negative integer $l < k$ and $\mathcal{E}' \in \Gamma_l(\mathcal{P})$ such that $\mathcal{E}' \sqsubseteq_{ms} \mathcal{E}$, and there is no $\mathcal{E}' \in \Gamma_k(\mathcal{P})$ such that $\mathcal{E}' \sqsubset_{ms} \mathcal{E}\}$.*

The following example shows how to compute the set of $\sqsubseteq_{ms}^1$-explanations for a BCQ by continuing our running example.

Example 2. Consider the BCQ $Q = \exists y : \mathsf{Person}(\mathsf{Tom}) \land \mathsf{worksFor}(\mathsf{Tom}, y)$ and the corresponding instance of the query abduction problem $\mathcal{P} = (\mathcal{T}, \mathcal{A}, Q, \Sigma)$, where $\mathcal{T}$ and $\mathcal{A}$ are given in Example 1 and $\Sigma = \{\mathsf{Person}, \mathsf{Employee}, \mathsf{worksFor}\}$. We can compute in turn that $\Xi_0(\mathcal{P}) = \emptyset$, $\Gamma_0(\mathcal{P}) = \emptyset$, $\Xi_1(\mathcal{P}) = \{\mathsf{Employee}(\mathsf{Tom}),$ $\mathsf{worksFor}(\mathsf{Tom}, y)\}$, and $\Gamma_1(\mathcal{P}) = \{\mathsf{Employee}(\mathsf{Tom}), \mathsf{worksFor}(\mathsf{Tom}, u)\}$, where y is a variable and u is a fresh individual. Hence $\sqsubseteq_{ms}^k$-expl($\mathcal{P}$) $\doteq \{\mathsf{Employee}(\mathsf{Tom}),$ $\mathsf{worksFor}(\mathsf{Tom}, u)\}$.

The representative explanations cannot be computed in an equally efficient way, because the verification of a representative explanation needs to consider representative explanations with larger cardinalities. For example, given two minimal explanations $\mathcal{E} = \{r(x, x)\}$ and $\mathcal{E}' = \{r(x, y), r(y, z)\}$, we cannot check whether $\mathcal{E}$ is representative before $\mathcal{E}'$ is generated. In fact, $\mathcal{E}$ is not representative because $\mathcal{E}' \cdot \{y \mapsto x, z \mapsto x\} = \mathcal{E}$ and there is no substitution θ such that $\mathcal{E}\theta \subseteq \mathcal{E}'$.

4   The More-Answers Problem

As mentioned before, users of CQA may often confront a problem that there are insufficiently many or even no certain answers to a given CQ. In this situation, users may want to see more answers with their corresponding explanations to tell why they are not certain answers. To fulfill this requirement, we generalize the CQA problem by making use of $\sqsubset_{ms}$-explanations and obtain a new problem for query answering, called the *more-answers* problem and defined below.

Definition 2 (the more-answers problem). *Given a consistent DL ontology* $\mathcal{O} = \mathcal{T} \cup \mathcal{A}$, *a CQ* $Q(\overrightarrow{x})$, *a set* Σ *of abducible predicates and a set* S *of certain or negative answers to* $Q(\overrightarrow{x})$ *in* $\mathcal{O}$, *we call* $\mathcal{M} = (\mathcal{T}, \mathcal{A}, Q(\overrightarrow{x}), \Sigma, S)$ *an instance of the* more-answers *problem. We define an* extra k-answer $\overrightarrow{t}$ *for* $\mathcal{M}$ *as a tuple of individuals occurring in* $\mathcal{T} \cup \mathcal{A}$ *or* $Q(\overrightarrow{x})$ *such that* $\overrightarrow{t} \notin S$ *and* $\sqsubset_{ms}^{k}$-$\mathsf{expl}(\mathcal{P}) \neq \emptyset$ *for* $\mathcal{P} = (\mathcal{T}, \mathcal{A}, Q(\overrightarrow{t}), \Sigma)$. *Let* $\mathsf{extans}(\mathcal{M}, k)$ *denote the set of extra* k-answers *for* $\mathcal{M}$, *and* $\mathsf{extans}^*(\mathcal{M})$ *denote the first nonempty* $\mathsf{extans}(\mathcal{M}, k)$ *for* k *increasing from 0 step by step. The goal for* $\mathcal{M}$ *is computing* $\mathsf{extans}^*(\mathcal{M})$.

Note that $\mathsf{extans}(\mathcal{M}, 0)$ for $\mathcal{M} = (\mathcal{T}, \mathcal{A}, Q(\overrightarrow{x}), \Sigma, \emptyset)$ is the set of certain answers to $Q(\overrightarrow{x})$ in $\mathcal{T} \cup \mathcal{A}$. Hence the more-answers problem generalizes the problem of computing $\mathsf{ans}(\mathcal{T} \cup \mathcal{A}, Q(\overrightarrow{x}))$, namely the traditional CQA problem.

By adapting the method given in Theorem 1, we can obtain a level-wise method for computing $\mathsf{extans}^*(M)$. Let $\Xi_k'(\mathcal{T}, \mathcal{A}, Q(\overrightarrow{x}), \Sigma) = \{\langle Q_1(\overrightarrow{x_1}\theta), \theta\rangle \mid Q'(\overrightarrow{x}) \in \tau(Q(\overrightarrow{x}), \mathcal{T}_D), \langle Q_1(\overrightarrow{x_1}), Q_2(\overrightarrow{x_2})\rangle$ is a bipartition of $Q'(\overrightarrow{x})$ such that $|Q_1(\overrightarrow{x_1})| = k$ and $Q_1(\overrightarrow{x_1})$ contains only predicates in Σ, and θ is a ground substitution for $Q_2(\overrightarrow{x_2})$ such that $Q_2(\overrightarrow{x_2}\theta) \subseteq \mathcal{A}\}$, $\Gamma_k'(\mathcal{T}, \mathcal{A}, Q(\overrightarrow{x}), \Sigma) = \{\langle E\sigma, \theta\sigma\rangle \mid \langle E, \theta\rangle \in \Xi_k'(\mathcal{T}, \mathcal{A}, Q(\overrightarrow{x}), \Sigma)$ and σ is either an empty substitution or a variant substitution for E such that $|E\sigma| < |E|\}$, and $\Phi_k(\mathcal{T}, \mathcal{A}, Q(\overrightarrow{x}), \Sigma) = \{\overrightarrow{x}\theta\eta \mid \langle E, \theta\rangle \in \Gamma_k'(\mathcal{T}, \mathcal{A}, Q(\overrightarrow{x}), \Sigma)$ and η is a ground substitution for $\overrightarrow{x}\theta$ such that all individuals in $\overrightarrow{x}\theta\eta$ appear in either $\mathcal{T} \cup \mathcal{A}$ or $Q(\overrightarrow{x})$ and $\mathcal{T} \cup \mathcal{A} \cup E\eta \not\models \bot\}$. The following theorem shows a sound and complete method for computing $\mathsf{extans}^*(M)$ using the above notations.

Theorem 2. $\mathsf{extans}^*(\mathcal{M}) = \Phi_k(\mathcal{T}, \mathcal{A}, Q(\overrightarrow{x}), \Sigma) \setminus S$ *where* k *is the minimum integer such that* $\Phi_k(\mathcal{T}, \mathcal{A}, Q(\overrightarrow{x}), \Sigma) \not\subseteq S$.

The corresponding $\sqsubset_{ms}^{k}$-explanations for an extra k-answer $\overrightarrow{t}$ in $\mathsf{extans}^*(\mathcal{M})$ can be computed accordingly. Let $\Psi_k(\mathcal{T}, \mathcal{A}, Q(\overrightarrow{t}), \Sigma) = \{E\eta \mid \langle E, \theta\rangle \in \Gamma_k'(\mathcal{T}, \mathcal{A}, Q(\overrightarrow{x}), \Sigma), \overrightarrow{x}\theta\eta = \overrightarrow{t}\}$, where variables in $\Psi_k(\mathcal{T}, \mathcal{A}, Q(\overrightarrow{t}), \Sigma)$ are treated as fresh individuals. By Theorems 1 and 2, we can obtain a sound and complete method for computing $\sqsubset_{ms}^{k}$-$\mathsf{expl}(\mathcal{P})$ from $\Psi_k(\mathcal{T}, \mathcal{A}, Q(\overrightarrow{t}), \Sigma)$ using the above notation, as shown in the following theorem.

Theorem 3. *Suppose* $\mathsf{extans}^*(\mathcal{M}) = \mathsf{extans}(\mathcal{M}, k)$. *Let* $\overrightarrow{t}$ *be an arbitrary extra* k-answer *for* $\mathcal{M}$, *then* $\sqsubset_{ms}^{k}$-$\mathsf{expl}(\mathcal{T}, \mathcal{A}, Q(\overrightarrow{t}), \Sigma) \doteq \{\mathcal{E} \in \Psi_k(\mathcal{T}, \mathcal{A}, Q(\overrightarrow{t}), \Sigma) \mid$ *there is no* $\mathcal{E}' \in \Psi_k(\mathcal{T}, \mathcal{A}, Q(\overrightarrow{t}), \Sigma)$ *such that* $\mathcal{E}' \sqsubset_{ms} \mathcal{E}\}$, *where* $\sqsubset_{ms}^{k}$-$\mathsf{expl}(\mathcal{T}, \mathcal{A}, Q(\overrightarrow{t}), \Sigma)$ *stands for* $\sqsubset_{ms}^{k}$-$\mathsf{expl}(\mathcal{P})$ *for* $\mathcal{P} = (\mathcal{T}, \mathcal{A}, Q(\overrightarrow{t}), \Sigma)$.

Consider the time complexity for computing all extra k-answers for $\mathcal{M}$ and their corresponding $\sqsubset_{ms}^{k}$-explanations in terms of data complexity. Since the number of CQs in $\tau(Q(\overrightarrow{x}), \mathcal{T}_D)$ and the number of bipartitions of a CQ in $\tau(Q(\overrightarrow{x}), \mathcal{T}_D)$

are independent from $\mathcal{A}$, $\Xi'_k(\mathcal{T}, \mathcal{A}, Q(\overrightarrow{x}), \Sigma)$ and $\Gamma'_k(\mathcal{T}, \mathcal{A}, Q(\overrightarrow{x}), \Sigma)$ can be computed in time polynomial in $|\mathcal{A}|$ under an assumption that the size of $\mathcal{T}$ and the size of $Q(\overrightarrow{x})$ are constants. Moreover, since $\mathcal{T} \cup \mathcal{A} \cup E\eta \not\models \bot$ if and only if there is no $Q' \in \gamma(\mathcal{T}_C \cup \mathcal{T}_E, \mathcal{T}_D)$ such that $\mathcal{A} \cup E\eta \not\models Q'$, $\Phi_k(\mathcal{T}, \mathcal{A}, Q(\overrightarrow{x}), \Sigma)$ can also be computed in polynomial time. Hence both the methods given in Theorems 2 and 3 work in polynomial time in data complexity.

The following example shows how to solve an instance of the more-answers problem by continuing our running example.

Example 3. Let $\mathcal{M} = (\mathcal{T}, \mathcal{A}, Q(\langle x \rangle), \Sigma, \{\langle \mathsf{Amy} \rangle\})$ be an instance of the more-answers problem, where $\mathcal{T}$, $\mathcal{A}$ and $Q(\langle x \rangle)$ are given in Example 1 and Σ is given in Example 2. We can compute that $\Xi'_0(\mathcal{T}, \mathcal{A}, Q(\langle x \rangle), \Sigma) = \{\langle \emptyset, \{x \mapsto \mathsf{Amy}, y \mapsto \mathsf{IBM}\}\rangle\}$, $\Gamma'_0(\mathcal{T}, \mathcal{A}, Q(\langle x \rangle), \Sigma) = \Xi'_0(\mathcal{T}, \mathcal{A}, Q(\langle x \rangle), \Sigma)$, and $\Phi_0(\mathcal{T}, \mathcal{A}, Q(\langle x \rangle), \Sigma) = \{\langle \mathsf{Amy} \rangle\}$. Since $\Phi_0(\mathcal{T}, \mathcal{A}, Q(\langle x \rangle), \Sigma) \subseteq S$, we need to continue level 1. We then compute that $\Xi'_1(\mathcal{T}, \mathcal{A}, Q(\langle x \rangle), \Sigma) = \{\langle \{\mathsf{Employee}(x)\}, \emptyset \rangle, \langle \{\mathsf{worksFor}(x, y)\}, \emptyset \rangle\}$, $\Gamma'_1(\mathcal{T}, \mathcal{A}, Q(\langle x \rangle), \Sigma) = \Xi'_1(\mathcal{T}, \mathcal{A}, Q(\langle x \rangle), \Sigma)$, and $\Phi_1(\mathcal{T}, \mathcal{A}, Q(\langle x \rangle), \Sigma) = \{\langle \mathsf{Amy} \rangle, \langle \mathsf{Tom} \rangle\}$. Since $\Phi_1(\mathcal{T}, \mathcal{A}, Q(\langle x \rangle), \Sigma) \not\subseteq S$, we have $\mathsf{extans}^*(\mathcal{M}) = \Phi_1(\mathcal{T}, \mathcal{A}, Q(\langle x \rangle), \Sigma) \setminus S = \{\langle \mathsf{Tom} \rangle\}$. Finally, we can compute the corresponding $\sqsubset^1_{\mathsf{ms}}$-explanations for the unique answer $\langle \mathsf{Tom} \rangle$ in $\mathsf{extans}^*(\mathcal{M})$. We compute that $\Psi_1(\mathcal{T}, \mathcal{A}, Q(\langle \mathsf{Tom} \rangle), \Sigma) = \{\{\mathsf{Employee}(\mathsf{Tom})\}, \{\mathsf{worksFor}(\mathsf{Tom}, u)\}\}$, where u is a fresh individual. Hence $\sqsubset^1_{\mathsf{ms}}\text{-}\mathsf{expl}(\mathcal{T}, \mathcal{A}, Q(\langle \mathsf{Tom} \rangle), \Sigma) \doteq \{\{\mathsf{Employee}(\mathsf{Tom})\}, \{\mathsf{worksFor}(\mathsf{Tom}, u)\}\}$.

5   The Proposed System

We developed a system for incrementally computing answers that have explanations in a user interactive mode (see http://www.dataminingcenter.net/cqa/ for more details about the system). The system was implemented in Java, using the Requiem [10] API for query rewriting and the MySQL engine to store and access ABoxes. Note that certain answers have explanations that are empty sets, thus they are also computed by our system. In the system, certain and negative answers are incrementally computed by solving a sequence of instances of the more-answers problem. In more details, given a first-order rewritable TBox $\mathcal{T}$, an ABox $\mathcal{A}$ and a CQ $Q(\overrightarrow{x})$, the system works through the following steps.

step 1: $S \leftarrow \mathsf{extans}^*(\mathcal{M})$ for $\mathcal{M} = (\mathcal{T}, \mathcal{A}, Q(\overrightarrow{x}), \emptyset, \emptyset)$;
step 2: Once users specify the set Σ of abducible predicates and request more answers, $S' \leftarrow S \cup \mathsf{extans}^*(\mathcal{M})$ for $\mathcal{M} = (\mathcal{T}, \mathcal{A}, Q(\overrightarrow{x}), \Sigma, S)$;
step 3: If $S' \supset S$, then $S \leftarrow S'$ and go to step 2.

While the system computes $\mathsf{extans}^*(\mathcal{M})$, it also computes the corresponding $\sqsubset_{\mathsf{ms}}$-explanations for every answer in $\mathsf{extans}^*(\mathcal{M})$. In fact, the system computes answers in a background thread and immediately outputs an answer and one of its explanations once the answer is worked out. In this way users can see answers as soon as possible. The output explanation may not be a $\sqsubset_{\mathsf{ms}}$-explanation initially, but it will be changed to a $\sqsubset_{\mathsf{ms}}$-explanation after the computation of $\mathsf{extans}^*(\mathcal{M})$ is done. Users can stop the computation whenever they do not want to see more answers.

The following example shows how the system works by continuing our running example.

Example 4. Suppose a user loads the ontology $\mathcal{O} = \mathcal{T} \cup \mathcal{A}$ and the CQ $Q(\langle x \rangle)$ given in Example 1 into the proposed system. Initially, the system computes extans*$(\mathcal{M})$ for $\mathcal{M} = (\mathcal{T}, \mathcal{A}, Q(\overrightarrow{x}), \emptyset, \emptyset)$. Suppose the user does not stop the computation, then the system outputs $S_1 = \{\langle \mathsf{Amy} \rangle\}$ with a $\sqsubseteq_{ms}$-explanation which is an empty set. If the user specifies the set Σ of abducible predicates as that given in Example 2 and requests more answers, the system will compute extans*$(\mathcal{M})$ for $\mathcal{M} = (\mathcal{T}, \mathcal{A}, Q(\overrightarrow{x}), \Sigma, S_1)$. Suppose again the user does not stop the computation, then the system outputs $S_2 = \{\langle \mathsf{Amy} \rangle, \langle \mathsf{Tom} \rangle\}$ with two $\sqsubseteq_{ms}$-explanations respectively corresponding to the two answers. Since $S_2 \supset S_1$, the user can request more answers. Suppose the user specifies the same set of abducible predicates and requests more answers again, then the system computes extans*$(\mathcal{M})$ for $\mathcal{M} = (\mathcal{T}, \mathcal{A}, Q(\overrightarrow{x}), \Sigma, S_2)$. This will not result in any new answer outside S_2. Hence the system finishes computing answers that have explanations and disables users to request more answers.

6   Experimental Evaluation

We conducted experiments on the Lehigh University Benchmark (LUBM) [9] to verify the efficiency and scalability of the proposed system. The LUBM TBox has been almost first-order rewritable and we only need to remove a few axioms that the Requiem API cannot handle and make it first-order rewritable. The slightly modified LUBM TBox has 43 concept names, 32 role names and 88 axioms. By using the LUBM generator available at http://swat.cse.lehigh.edu/projects/lubm/, we created four LUBM ontologies with large ABoxes, written LUBMn, where $n = 1, 5, 10, 50$ denotes the number of universities. The number of individuals in LUBMn is from 17,174 to 1,082,818, while the number of assertions in the ABox of LUBMn is from 100,543 to 6,863,227.

We verified how well the system works for the first two rounds of computation. In the first round, the system computes $S = $ extans*$(\mathcal{M})$ for $\mathcal{M} = (\mathcal{T}, \mathcal{A}, Q(\overrightarrow{x}), \emptyset, \emptyset)$. The resulting set S amounts to the set of ceratin answers to $Q(\overrightarrow{x})$ in $\mathcal{T} \cup \mathcal{A}$. In the second round, the system computes extans*$(\mathcal{M})$ for $\mathcal{M} = (\mathcal{T}, \mathcal{A}, Q(\overrightarrow{x}), \Sigma, S)$, where Σ was specified as the set of concept names and role names. The given CQ $Q(\overrightarrow{x})$ was set as any of the 14 benchmark CQs provided in [9], with the first variable treated as the unique answer variable and remaining variables treated as quantified variables. A time limit of one hour was set for evaluating an individual CQ in either round. All our experiments were conducted on a laptop with Intel Dual-Core 2.20 GHz CPU and 4 GB RAM, running Windows 7, where the maximum Java heap size was set to 3 GB.

The experimental results for the first round are reported in Table 1, where the execution time in this round includes the time for rewriting the given CQ, the time for retrieving certain answers from the ABox and the time for displaying all retrieved answers. This round amounts to performing traditional CQA through

Table 1. The execution time and the number of computed answers in the first round

CQ	LUBM1		LUBM5		LUBM10		LUBM50	
	Time	#Ans	Time	#Ans	Time	#Ans	Time	#Ans
Q_1	37	4	37	4	37	4	37	4
Q_2	36	0	106	9	158	28	1,733	130
Q_3	39	6	39	6	42	6	42	6
Q_4	125	34	138	34	136	0	167	0
Q_5	194	719	205	719	205	719	223	719
Q_6	3,091	7,790	37,767	48,582	152,169	99,566	3,223,262	519,842
Q_7	711	66	745	66	982	66	989	66
Q_8	2,900	7,790	2,977	7,790	605	0	641	0.
Q_9	2,138,942	207	2,156,091	1,239	2,173,428	2,514	2,460,647	13,513
Q_{10}	149	4	153	4	155	4	198	4
Q_{11}	35	0	36	0	39	0	40	0
Q_{12}	94	15	99	15	111	15	146	15
Q_{13}	79	1	86	21	127	33	152	228
Q_{14}	1,303	5,916	18,894	36,682	449,925	75,547	1,679,140	393,730

Note: Time is the execution time in milliseconds; #Ans is the number of certain answers computed.

query rewriting. For almost all benchmark CQs except Q_6, Q_9 and Q_{14}, the proposed system works efficiently and scales well to millions of assertions. The CQs Q_6 and Q_{14} are *atomic queries* that are composed of single atoms, hence the number of certain answers increases heavily with the number of assertions, and so does the execution time for displaying answers. The CQ Q_9 requires a long time for rewriting (this might be an issue of Requiem [10]), thus the reported execute time is rather long.

The experimental results for the second round are reported in Table 2, where the execution time in this round includes the time for computing negative answers and their corresponding explanations and the time for displaying them. For CQs Q_3, Q_5, Q_{13} and Q_{14}, the system works in milliseconds even with millions of assertions. For other CQs, the execution time is roughly in direct proportion to the answers computed. It also smoothly increases with the number of assertions except for CQs Q_4 and Q_8 where the number of answers computed drastically drops when the test ontology is changed from LUBM5 to LUBM10. The execution time in this round is longer than that in the first round except for Q_5, Q_9 and Q_{14}, mainly due to more answers to be computed. The reason for Q_9 is that this round does not require query rewriting. The reason for Q_5 and Q_{14} is that this round only computes one new answer and finishes quickly.

At a word, the above experimental results show that the proposed system works well for general CQs that are not atomic. The problem for atomic queries is that there will possibly be too many answers to display. We believe that this

Table 2. The execution time and the number of computed answers in the second round

CQ	LUBM1		LUBM5		LUBM10		LUBM50	
	Time	#Ans	Time	#Ans	Time	#Ans	Time	#Ans
Q_1	5,320	1,874	50,947	11,900	248,652	24,019	421,541	126,112
Q_2	5,540	1,875	35,332	11,903	187,015	24,024	688,781	126,144
Q_3	40	7	70	7	83	7	152	7
Q_4	29,471	8,331	150,610	51,956	2,342	675	2,607	675
Q_5	96	720	143	720	166	720	197	720
Q_6	2,285,643	8,331	Timeout	48,583	Timeout	99,567	Timeout	519,843
Q_7	80,803	8,331	507,290	51,956	1,797,237	106,410	Timeout	368,014
Q_8	1,132,818	8,331	ROM	39,097	44,058	8,428	46,386	8,428
Q_9	32,639	7,791	198,655	48,583	490,004	99,567	Timeout	508,129
Q_{10}	39,708	8,331	176,287	51,956	825,016	106,410	Timeout	519,843
Q_{11}	725	239	3,437	1,386	22,849	2,842	65,074	14,876
Q_{12}	1,529	541	3,943	619	8,020	715	37,967	1,525
Q_{13}	81	2	139	22	237	34	327	229
Q_{14}	36	5,917	81	36,683	105	75,548	191	393,731

Note: Time is the execution time in milliseconds, where ROM is short for running out of memory; #Ans is the total number of answers computed after finish, timeout or running out of memory (including certain answers computed in the first round).

problem can be solved by exploiting some optimization techniques to display computed answers. On the other hand, the system shows a good scalability against increasing number of assertions. It scales up to millions of assertions for most benchmark CQs.

7   Conclusion and Future Work

In this paper we have addressed the problem of providing users negative answers to a given conjunctive query in a consistent DL ontology while explaining why these answers are not certain ones. We first proposed a new class of explanations, $\sqsubseteq_{ms}$-explanations, for certain and negative answers, then proposed a new problem for query answering, the more-answers problem, in order to incrementally compute certain and negative answers that have explanations. Targeting consistent DL ontology whose TBox is first-order rewritable, we proposed a tractable method (in data complexity) for solving the more-answers problem. We also proposed a system that incrementally computes certain and negative answers by solving a sequence of instances of the more-answers problem. The system was empirically shown to be scalable for DL ontologies with large ABoxes.

There are at least two directions that can be explored in the future work. On the one hand, more evaluations should be conducted to verify the proposed system in real-life applications. On the other hand, since inconsistency may often occur in real-life applications such as data integration and ontology population [6], it is important to define and compute reasonable answers and their corresponding explanations in an inconsistent DL ontology.

Acknowledgments. This work is partly supported by the NSFC grants (61375056 and 61005043), the Guangdong Natural Science Foundation (S2013010012928), the Undergraduate Innovative Experiment Projects in Guangdong University of Foreign Studies (1184613038, 1184613026 and 201411846043), and the Business Intelligence Key Team of Guangdong University of Foreign Studies (TD1202).

References

1. Baader, F., Calvanese, D., McGuinness, D.L., Nardi, D., Patel-Schneider, P.F. (eds.): The Description Logic Handbook: Theory, Implementation, and Applications. Cambridge University Press, Cambridge (2003)
2. Borgida, A., Calvanese, D., Rodriguez-Muro, M.: Explanation in the $DL - Lite$ family of description logics. In: Meersman, R., Tari, Z. (eds.) OTM 2008, Part II. LNCS, vol. 5332, pp. 1440–1457. Springer, Heidelberg (2008)
3. Calì, A., Gottlob, G., Lukasiewicz, T.: A general datalog-based framework for tractable query answering over ontologies. J. Web Semant. **14**, 57–83 (2012)
4. Calvanese, D., Giacomo, G., Lembo, D., Lenzerini, M., Rosati, R.: Tractable reasoning and efficient query answering in description logics: the DL-Lite family. J. Autom. Reasoning **39**(3), 385–429 (2007)
5. Calvanese, D., Ortiz, M., Simkus, M., Stefanoni, G.: Reasoning about explanations for negative query answers in DL-Lite. J. Artif. Intell. Res. **48**, 635–669 (2013)
6. Du, J., Qi, G., Shen, Y.: Weight-based consistent query answering over inconsistent $\mathcal{SHIQ}$ knowledge bases. Knowl. Inf. Syst. **34**(2), 335–371 (2013)
7. Du, J., Wang, K., Shen, Y.: A tractable approach to ABox abduction over description logic ontologies. In: Proceedings of the 28th AAAI Conference on Artificial Intelligence (AAAI), pp. 1034–1040 (2014)
8. Du, J., Wang, S., Qi, G., Pan, J.Z., Hu, Y.: A new matchmaking approach based on abductive conjunctive query answering. In: Pan, J.Z., Chen, H., Kim, H.-G., Li, J., Wu, Z., Horrocks, I., Mizoguchi, R., Wu, Z. (eds.) JIST 2011. LNCS, vol. 7185, pp. 144–159. Springer, Heidelberg (2012)
9. Guo, Y., Pan, Z., Heflin, J.: LUBM: a benchmark for OWL knowledge base systems. J. Web Semant. **3**(2–3), 158–182 (2005)
10. Pérez-Urbina, H., Motik, B., Horrocks, I.: Tractable query answering and rewriting under description logic constraints. J. Appl. Logic **8**(2), 186–209 (2010)

The *Ontop* Framework for Ontology Based Data Access

Timea Bagosi[1], Diego Calvanese[1], Josef Hardi[2], Sarah Komla-Ebri[1],
Davide Lanti[1], Martin Rezk[1], Mariano Rodríguez-Muro[3], Mindaugas Slusnys[1],
and Guohui Xiao[1](✉)

[1] Faculty of Computer Science, Free University of Bozen-Bolzano,
Bolzano, Italy
xiao@inf.unibz.it
[2] Obidea Technology, Jakarta, Indonesia
[3] IBM T.J. Watson Research Center,
Yorktown Heights, NY, USA

1 Ontology Based Data Access

Ontology Based Data Access (OBDA) [4] is a paradigm of accessing data trough a conceptual layer. Usually, the conceptual layer is expressed in the form of an RDF(S) [10] or OWL [15] ontology, and the data is stored in relational databases. The terms in the conceptual layer are mapped to the data layer using mappings which associate to each element of the conceptual layer a (possibly complex SQL) query over the data sources. The mappings have been formalized in the recent R2RML W3C standard [6]. This virtual graph can then be queried using an RDF query language such as SPARQL [7].

Formally, an OBDA system is a triple $\mathcal{O} = \langle \mathcal{T}, \mathcal{S}, \mathcal{M} \rangle$, where:

- $\mathcal{T}$ is the intensional level of an ontology. We consider ontologies formalized in description logics (DLs), hence $\mathcal{T}$ is a DL TBox.
- $\mathcal{S}$ is a relational database representing the sources.
- $\mathcal{M}$ is a set of mapping assertions, each one of the form

$$\Phi(\boldsymbol{x}) \ \leftarrow \ \Psi(\boldsymbol{x})$$

where
- $\Phi(\boldsymbol{x})$ is a query over $\mathcal{S}$, returning tuples of values for $\boldsymbol{x}$
- $\Psi(\boldsymbol{x})$ is a query over $\mathcal{T}$ whose free variables are from $\boldsymbol{x}$.

The main functionality of OBDA systems is query answering. A schematic description of the query transformation process (usually SPARQL to SQL) performed by a typical OBDA system is provided in Fig. 1. In such an architecture, queries posed over a conceptual layer are translated into a query language that can be handled by the data layer. The translation is independent of the actual data in the data layer. In this way, the actual query evaluation can be delegated to the system managing the data sources.

© Springer-Verlag Berlin Heidelberg 2014
D. Zhao et al. (Eds.): CSWS 2014, CCIS 480, pp. 67–77, 2014.
DOI: 10.1007/978-3-662-45495-4_6

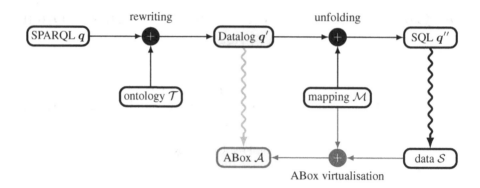

Fig. 1. Query processing in an OBDA system

2   The *Ontop* Framework

Ontop is an open-source OBDA framework released under the Apache license, developed at the Free University of Bozen-Bolzano[1] and currently acts as the query transformation module of the EU project Optique[2].

As an OBDA system, to the best of our knowledge, *Ontop* is the first to support all the following W3C recommendations: OWL, R2RML, SPARQL, SWRL and SPARQL OWL 2 QL regime. In addition, all the major commercial and free databases are supported. For each component of the OBDA system, *Ontop* supports the widely used standards:

Mapping. *Ontop* supports two mapping languages: (1) the native *Ontop* mapping language which is easy to learn and use and (2) the RDB2RDF Mapping Language (R2RML) which is a W3C recommendation.

Ontology. *Ontop* fully supports OWL 2 QL ontology language [11], which is a superset of RDFS. OWL 2 QL is based on the DL-Lite family of description logics [5], which are lightweight ontologies and guarantee queries over the ontology can be rewritten to equivalent queries over the data source. Recently *Ontop* is also extended to support the linear recursive fragment of SWRL (Semantic Web Rule Language) [8,16].

Data Source. *Ontop* supports all the databases which implement SQL 99. These include all major relational database systems, e.g., PostgreSQL, MySQL, H2, DB2, ORACLE, and MS SQL Server.

Query. *Ontop* essentially supports all the features of SPARQL 1.0 and SPARQL OWL QL Regime of SPARQL 1.1 [9]. Supporting of other features in SPARQL 1.1 (e.g., aggregates, property path queries, negations) is ongoing work.

---

[1] http://ontop.inf.unibz.it
[2] http://www.optique-project.eu

The core of the *Ontop* is the SPARQL engine Quest which supports RDFS and OWL 2 QL entailment regimes by rewriting the SPARQL queries (over the virtual RDF graph) to SQL queries (over the relational database). *Ontop* is able to generate efficient (and highly optimized [13,14]) SQL queries, that in some cases are very close to the SQL queries that would be written by a database expert.

The *Ontop* framework can be used as:

- a *plugin for Protege 4* which provides a graphical interface for mapping editing and SPARQL query execution,
- a *Java library* which implements both OWL API and Sesame API interfaces, available as maven dependencies, and
- a *SPARQL end-point* through Sesame's Workbench.

3   A Demo of the Movie Scenario

In this section, we describe a complete demo of *Ontop* using the movie scenario [12]. The datasets and systems are available online[3].

3.1   Movie Scenario Dataset

The Movie Ontology. The movie ontology *MO* aims to provide a controlled vocabulary to semantically describe movie related concepts (e.g., Movie, Genre, Director, Actor) and the corresponding individuals ("Ice Age", "Drama", "Steven Spielberg" or "Johnny Depp") [3]. The ontology contains concept hierarchies for movie categorization that enables user-friendly presentation of movie descriptions in the appropriate detail. There are several additions to the ontology terminology due to the requirements in the demo, e.g., concepts TVSeries and Actress.

IMDb Data. IMDB's data is provided as text files[4] which need to be converted into an SQL file using a third party tool. Our IMDB raw data was downloaded in 2010 and the SQL script was generated using IMDbPY[5]. IMDbPY generates an SQL schema (tables) appropriate for storing IMDB data and then reads the IMDB plain text data files to generate the SQL INSERT commands that populate the tables. It can generate PostgreSQL, MySQL and DB2 SQL scripts. In this demo we use a PostgreSQL compatible script and database takes up around 6 GB on the disk.

Mappings. The mappings for this scenario are natural mappings that associate the data in the SQL database to the movie ontology's vocabulary. They are "natural" mapping, in the sense that the only purpose of the mappings was

---

[3] https://github.com/ontop/ontop/wiki/Example_MovieOntology
[4] http://www.imdb.com/interfaces
[5] http://imdbpy.sourceforge.net

to be able to query the data through the ontology. There was no intention to highlight the benefits of any algorithm or technique used in *Ontop*. The first version of the mappings for this scenario were developed by students of Free University of Bolzano as part of an lab assignment. The current mappings are the improved version of those create by our development team.

Queries. We included around 40 queries which are in the file `movieontology.q` and can be used to explore the data set. The queries have different complexities, going from very simple to fairly complex. Note that some form of inference (beyond simple query evaluation) is involved in most of these queries, in particular, hierarchies are often involved.

3.2 Using Protege Plugin

We demonstrate how to use *Ontop* as a protege plugin. The steps are:

(1) Start PostgreSQL with IMDb data.
(2) Start Protege with ontop plugin from command line.

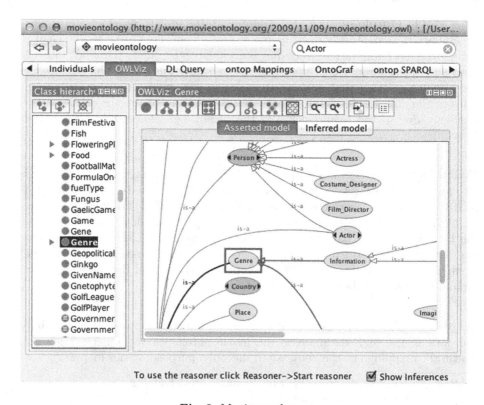

Fig. 2. Movie ontology

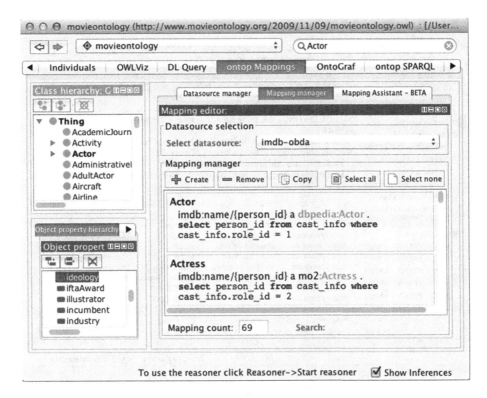

Fig. 3. Movie mappings

(3) Open the OWL file `movieontology.owl` from Protege. The *Ontop* plugin will also automatically open the mapping file `movieontology.obda` and query file `movieontology.q`.

(4) Check the ontology and mappings. Two screen shots of the ontology and mappings are shown in Figs. 2 and 3.

(5) Start the Quest reasoner from the menu.

(6) Run sample queries and check the generated SQLs. For example, we can execute the query "Find names that act as both the director and the actor at the same time produced in Eastern Asia" as shown in Fig. 4.

3.3   Using Java API

We show how the movie scenario can be implemented using the *Ontop* java libraries through OWL API and sesame API. The complete code for the demo is available online[6].

Using OWL API. The OWL API is a Java API and reference implementation for creating, manipulating and serializing OWL Ontologies [2]. In the first

---

[6] https://github.com/ontop/ontop-examples

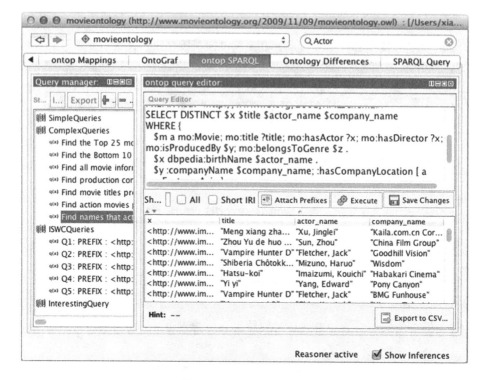

Fig. 4. Example query

example we use OWL API to execute all the 40 SPARQL queries over the movie ontology, using the mapping in our obda format and a PostgreSQL database with the IMDb data.

Ontop uses Maven to manage the dependencies. Since the release of version 1.10, *Ontop* itself has been deployed to the central maven repository. All artifacts have the same groupId it.unibz.inf.ontop. In this example we use the OWL API interface of *Ontop*, so we put the following in the pom.xml:

```
<dependency>
    <groupId>it.unibz.inf.ontop</groupId>
    <artifactId>ontop-quest-owlapi3</artifactId>
    <version>1.12.0</version>
</dependency>
```

Moreover we need the dependency for PostgreSQL JDBC driver as shown below.

```
<dependency>
    <groupId>postgresql</groupId>
    <artifactId>postgresql</artifactId>
    <version>9.0-801.jdbc4</version>
</dependency>
```

The files needed to start the *Ontop* reasoner are the ontology file `movieontology.owl` and the obda file `movieontology.obda`. The obda file contains both mappings and database settings. This allows to access the data in the PostgreSQL database using the mappings in the OBDA model. First we load the OWL file and OBDA file:

```
// Loading the OWL file using OWL API
OWLOntologyManager manager;
manager = OWLManager.createOWLOntologyManager();
OWLOntology ontology;
ontology = manager.loadOntologyFromOntologyDocument
        ((new File(owlFile)));

// Loading the OBDA file
OBDAModel obdaModel = fac.getOBDAModel();
ModelIOManager ioManager = new ModelIOManager(obdaModel);
ioManager.load(obdaFile);
```

Next we create a new instance of the reasoner (QuestOWL reasoner), adding the necessary preferences to prepare its configuration. We prepare the

```
QuestOWLFactory factory = new QuestOWLFactory();
factory.setOBDAController(obdaModel);

//Setting preferences putting Quest in virtual mode.
factory.setPreferenceHolder(p);
QuestPreferences preference = new QuestPreferences();
preference.setCurrentValueOf
    (QuestPreferences.ABOX_MODE, QuestConstants.VIRTUAL);

// Creating a new instance of the reasoner
QuestOWL reasoner;
reasoner = (QuestOWL) factory.createReasoner
        (ontology, new SimpleConfiguration());

// Now we are ready for querying
QuestOWLConnection conn = reasoner.getConnection();
QuestOWLStatement st = conn.createStatement();
```

Ontop supports a file format of multiple SPARQL queries. Here we execute each query using the file `movieontology.q` of 40 queries. Within the instance each SPARQL query is translated in an SQL query, which allows to retrieve the results from the PostgreSQL database. For simplicity, we only display to the user the number of results of the query and the time required for the execution.

```
// Loading the query file
QueryController qc = new QueryController();
QueryIOManager qman = new QueryIOManager(qc);
qman.load("src/main/resources/example/movie/movieontology.q");

// Execute each query
for (QueryControllerGroup group : qc.getGroups()) {
    for (QueryControllerQuery query : group.getQueries()) {

        System.out.println("Executing_query:_" + query.getID());
        System.out.println("Query:_\n" + query.getQuery());

        long start = System.nanoTime();
        QuestOWLResultSet res = st.executeTuple(query.getQuery());
        long end = System.nanoTime();
        double time = (end - start) / 1000;
        int count = 0;
        while (res.nextRow()) {
            count += 1;
        }

        System.out.println("Total_result:_" + count);
        System.out.println("Elapsed_time:_" + time + "_ms");
    }
}
```

At the end of the execution we close all connections and we dispose of the reasoner.

```
//Close connection and resources
if (st != null && !st.isClosed()) {
    st.close();
}
if (!conn.isClosed()) {
    conn.close();
}
reasoner.dispose();
```

Using Sesame API. OpenRDF Sesame is a de-facto standard framework for processing RDF data and includes parsers, storage solutions (RDF databases a.ka. triplestores), reasoning and querying, using the SPARQL query language [1].

In the second example we show how to create a repository and execute a single query using Sesame API. First we need to add the Sesame API module of *Ontop* as a dependency to the pom file pom.xml.

```
<dependency>
  <groupId>it.unibz.inf.ontop</groupId>
```

```
<artifactId>ontop-quest-sesame</artifactId>
<version>1.12.0</version>
</dependency>
```

Then we set up the repository and create a connection. The repositories must always be initialized first. We get the repository connection that will be used to execute the query.

```
// Creating and initializing the repository
boolean existential = false;
String rewriting = "TreeWitness";
SesameVirtualRepo repo = new SesameVirtualRepo
        ("test_repo", owlFile, obdaFile, existential, rewriting);
repo.initialize();

RepositoryConnection conn = repo.getConnection();
```

We load the SPARQL file **q1Movie.rq** which contains the same query that we used for the Protege example.

```
//Loading the SPARQL file
String queryString = "";

BufferedReader br = new BufferedReader(new FileReader(sparqlFile));
String line;
while ((line = br.readLine()) != null) {
    queryString += line + "\n";
}
System.out.println();
System.out.println("The input SPARQL query:");
System.out.println("========================");
System.out.println(queryString);
System.out.println();
```

Now we are ready to execute the query using the created Sesame repository connection and output the results of the SPARQL from the database.

```
// Executing the query
Query query = conn.prepareQuery(QueryLanguage.SPARQL, queryString);

TupleQuery tq = (TupleQuery) query;

TupleQueryResult result = tq.evaluate();

while (result.hasNext()) {
    for (Binding binding : result.next()) {
            System.out.print(binding.getValue() + ", ");
    }
    System.out.println();
}
```

Finally we close all the connections and release the resources.

```
//Close result set to release resources
result.close();

// Finally close connection to release resources
System.out.println("Closing..,");
conn.close();
```

Acknowledgement. This paper is supported by the EU under the large-scale integrating project (IP) Optique (Scalable End-user Access to Big Data), grant agreement n. FP7-318338.

References

1. OpenRDF Sesame. http://www.openrdf.org/. Accessed 27 Aug 2014
2. Owl, API. http://owlapi.sourceforge.net/. Accessed 27 Aug 2014
3. Bouza, A.: MO - the movie ontology (2010). http://www.movieontology.org. Accessed 26 Jan 2010
4. Calvanese, D., De Giacomo, G., Lembo, D., Lenzerini, M., Poggi, A., Rodriguez-Muro, M., Rosati, R.: Ontologies and databases: the *DL-Lite* approach. In: Tessaris, S., Franconi, E., Eiter, T., Gutierrez, C., Handschuh, S., Rousset, M.-C., Schmidt, R.A. (eds.) Reasoning Web. LNCS, vol. 5689, pp. 255–356. Springer, Heidelberg (2009)
5. Calvanese, D., De Giacomo, G., Lembo, D., Lenzerini, M., Rosati, R.: Tractable reasoning and efficient query answering in description logics: the *DL-Lite* family. J. Autom. Reas. **39**(3), 385–429 (2007)
6. Das, S., Sundara, S., Cyganiak, R.: R2RML: RDB to RDF mapping language. W3C Recommendation, World Wide Web Consortium, September 2012. http://www.w3.org/TR/r2rml/
7. Harris, S., Seaborne, A.: SPARQL 1.1 Query Language. W3C Recommendation, World Wide Web Consortium, March 2013. http://www.w3.org/TR/sparql11-query
8. Horrocks, I., Patel-Schneider, P., Boley, H., Tabet, S., Grosof, B., Dean, M.: SWRL: a semantic web rule language combining OWL and RuleML. W3C Member Submission, World Wide Web Consortium (2004)
9. Kontchakov, R., Rezk, M., Rodríguez-Muro, M., Xiao, G., Zakharyaschev, M.: Answering SPARQL queries over databases under OWL 2 QL entailment regime. In: Mika, P., Tudorache, T., Bernstein, A., Welty, C., Knoblock, C., Vrandečić, D., Groth, P., Noy, N., Janowicz, K., Goble, C. (eds.) ISWC 2014, Part I. LNCS, vol. 8796, pp. 552–567. Springer, Heidelberg (2014)
10. Manola, F., Mille, E.: RDF primer. W3C Recommendation, World Wide Web Consortium, February 2004. http://www.w3.org/TR/rdf-primer-20040210/
11. Motik, B., Grau, B.C., Horrocks, I., Wu, Z., Fokoue, A., Lutz, C.: OWL 2 web ontology language: profiles. W3C Recommendation, World Wide Web Consortium (2012). http://www.w3.org/TR/owl2-profiles/

12. Rodriguez-Muro, M., Hardi, J., Calvanese, D.: Quest: efficient SPARQL-to-SQL for RDF and OWL. In: Glimm, B., Huynh, D. (eds.) International Semantic Web Conference (Posters & Demos). CEUR Workshop Proceedings, vol. 914. CEUR-WS.org (2012)
13. Rodríguez-Muro, M., Kontchakov, R., Zakharyaschev, M.: Ontology-based data access: *Ontop* of databases. In: Alani, H., Kagal, L., Fokoue, A., Groth, P., Biemann, C., Parreira, J.X., Aroyo, L., Noy, N., Welty, C., Janowicz, K. (eds.) ISWC 2013, Part I. LNCS, vol. 8218, pp. 558–573. Springer, Heidelberg (2013)
14. Rodriguez-Muro, M., Rezk, M., Hardi, J., Slusnys, M., Bagosi, T., Calvanese, D.: Evaluating SPARQL-to-SQL translation in Ontop. In: Proceedings of the 2nd International Workshop on OWL Reasoner Evaluation (ORE 2013). CEUR Workshop Proceedings, vol. 1015, pp. 94–100 (2013)
15. W3C OWL Working Group: OWL 2 web ontology language document overview (second edition). W3C Recommendation, World Wide Web Consortium (2012). http://www.w3.org/TR/owl2-overview/
16. Xiao, G., Rezk, M., Rodríguez-Muro, M., Calvanese, D.: Rules and ontology based data access. In: Kontchakov, R., Mugnier, M.-L. (eds.) RR 2014. LNCS, vol. 8741, pp. 157–172. Springer, Heidelberg (2014)

Complexity of Conservative Extensions and Inseparability in the Description Logic $\mathcal{EL}^\neg$

Yuming Shen[1]($\boxtimes$) and Ju Wang[2]

[1] Cisco School of Informatics, Guangdong University of Foreign Studies,
Guangzhou 510420, China
ymshen2002@163.com
[2] School of Computer Science and Information Engineering,
Guangxi Normal University, Guilin 541004, China

Abstract. The notations of conservative extensions and inseparability are suggested as the effective tool for comparing, merging, and modularizing description logic ontologies. It has been shown that the complexity of conservative extensions for expressive descriptions logics such as $\mathcal{ALC}$ and $\mathcal{ALCQI}$ are 2ExpTime-complete and ExpTime-complete for $\mathcal{EL}$ itself. However, the problem of the complexity of conservative extensions in a few extensions of $\mathcal{EL}$ which used in applications has hardly been addressed. The aim of this paper is to study the complexity of conservative extensions and inseparability in the description logic $\mathcal{EL}^\neg$, which is the extension of $\mathcal{EL}$ with atomic concept negation. By adding many countable new concept names which correspond to the complex negative concepts, we establish a translation from $\mathcal{ALC}$ to $\mathcal{EL}^\neg$ and reduce the problem of conservative extensions in $\mathcal{ALC}$ to the case of $\mathcal{EL}^\neg$. Since deciding conservative extensions and inseparability in $\mathcal{ALC}$ is 2ExpTime-complete, we get 2ExpTime-completeness of both inseparability and conservative extensions in $\mathcal{EL}^\neg$.

Keywords: Ontology · Conservative extension · Computational complexity

1 Introduction

The main use of ontologies in computer and information science is to provide a reference vocabulary for an application domain; that is, to give a logical theory which defines the semantics of the terms and relations in the vocabulary. In logic-based ontology languages such as description logics, the vocabulary is represented as concept, role and individual names. Notable examples of large ontologies in health care and the bio-sciences include the National Cancer Institute's (NCI) thesaurus [16], the gene ontology (GO) [18], and the Systematized Nomenclature of Medicine, Clinical Terms (SNOMED, CT), which comprises about 0.5 million vocabulary items [17]. In such an environment, automated tool support for the design, maintenance, refinement, and customization of large ontologies is critical importance.

© Springer-Verlag Berlin Heidelberg 2014
D. Zhao et al. (Eds.): CSWS 2014, CCIS 480, pp. 78–86, 2014.
DOI: 10.1007/978-3-662-45495-4_7

The notation of a conservative extension originates from mathematical logic where it is used for relative consistency proofs in arithmetic and set theory. Recently, the conservative extension and its generalization inseparability are suggested as the effective tool support for comparing, merging, modularizing, and reusing ontologies, e.g., [8–10]. In computer science, conservative extensions have found applications in modular software specification and verification, e.g., [3,14]. In answer set programming, modularity and variations of conservative extensions have been investigated in, e.g., [5,15].

Given two ontologies T_1 and T_2 with the condition $T_1 \subseteq T_2$, we say that T_2 is a Σ–conservative extension of T_1 if and only if they have the same logical consequences formulated in the signature (vocabulary) Σ. We say that T_1 and T_2 are Σ–inseparable if the two ontologies share the same logical consequence and the condition $T_1 \subseteq T_2$ is dropped. It has been shown that the complexity of conservative extensions investigated in [7,12] for expressive descriptions logics such as $\mathcal{ALC}$ and $\mathcal{ALCQI}$ are 2ExpTime-complete and ExpTime-complete for $\mathcal{EL}$ itself [13]. We note that the problem of the complexity of conservative extensions in a few extensions of $\mathcal{EL}$ which used in applications has hardly been addressed. The purpose of this paper is to help fill this gap. We study the complexity of conservative extensions and inseparability in the description logic $\mathcal{EL}^\neg$, which is the extension of $\mathcal{EL}$ with atomic concept negation. By adding many countable new concept names which correspond to the complex negative concepts, we establish a translation from $\mathcal{ALC}$ to $\mathcal{EL}^\neg$. The translation consists two parts: for every concept C in $\mathcal{ALC}$, $\sigma(C)$ is a concept in $\mathcal{EL}^\neg$, and for every TBox T in $\mathcal{ALC}$, $\sigma(T)$ is a TBox in $\mathcal{EL}^\neg$. Using the translation, we reduce the problem of conservative extensions in $\mathcal{ALC}$ to the case of $\mathcal{EL}^\neg$. Since deciding conservative extensions and inseparability in $\mathcal{ALC}$ is 2ExpTime-complete, we get 2ExpTime-completeness of both inseparability and conservative extensions in $\mathcal{EL}^\neg$.

This paper is organized as follows: the next section gives the preliminaries of the description logics $\mathcal{ALC}$ and $\mathcal{EL}^\neg$ used in the paper as well as the notations of conservative extensions and inseparability; the third section gives the translation from $\mathcal{ALC}$ to $\mathcal{EL}^\neg$ and some basic properties about the translation; the fourth section shows that deciding conservative extensions and inseparability in $\mathcal{ALC}$ can be reduced to the case of $\mathcal{EL}^\neg$, and these results establish that 2ExpTime-completeness of both inseparability and conservative extensions in $\mathcal{EL}^\neg$; the last section concludes the paper.

2   Preliminaries

In description logics, concept descriptions are inductively defined with the help of a set of constructors, staring with a set $\mathbf{N_C}$ of concept names, a set $\mathbf{N_R}$ of role names, and a set $\mathbf{N_I}$ of individual names. We introduce the syntax and semantics of the description logics considered in this paper, which are $\mathcal{ALC}$ and $\mathcal{EL}^\neg$; for a more comprehensive introduction, the reader is referred to the description logics handbook [1]. Let $\mathbf{N_C}$ and $\mathbf{N_R}$ be disjoint and countably infinite sets of concept names and role names. Then $\mathcal{ALC}$ concepts are built by the following syntax rules:

$$C :: = \top | A | \neg C | C \sqcap D | \exists r.C,$$

where A ranges over concept names and r ranges over role names. The concept constructors $\bot, \sqcup$, and $\forall r.C$ are defined as abbreviation: $\bot$ stands for $\neg \top$, $C \sqcup D$ for $\neg(\neg C \sqcap \neg D)$ and $\forall r.C$ abbreviates $\neg \exists r. \neg C$.

The representation of ontologies in description logics are based on TBoxes. Formally, an $\mathcal{ALC}-$TBox is a finite set of concept inclusions $C \sqsubseteq D$, where C and D are $\mathcal{ALC}-$concepts. We write $C \equiv D$ as an abbreviation for the two concept inclusions $C \sqsubseteq D$ and $D \sqsubseteq C$.

The semantics of $\mathcal{ALC}$ concepts is defined by means of interpretation $\mathcal{I} = (\Delta^{\mathcal{I}}, .^{\mathcal{I}})$, where the interpretation domain $\Delta^{\mathcal{I}}$ is a non-empty set, and $.^{\mathcal{I}}$ is the interpretation function, assigning each concept name A to a subset $A^{\mathcal{I}}$ of $\Delta^{\mathcal{I}}$, each role name r to a binary relation $r^{\mathcal{I}} \subseteq \Delta^{\mathcal{I}} \times \Delta^{\mathcal{I}}$, and each individual name a to an element $a^{\mathcal{I}} \in \Delta^{\mathcal{I}}$. The interpretation function is inductively extended to concepts as follows:

$$\top^{\mathcal{I}} = \Delta^{\mathcal{I}}$$
$$(\neg C)^{\mathcal{I}} = \Delta / C^{\mathcal{I}}$$
$$(C \sqcap D)^{\mathcal{I}} = C^{\mathcal{I}} \cap D^{\mathcal{I}}$$
$$(\exists r.C)^{\mathcal{I}} = \{x \in \Delta^{\mathcal{I}} \mid \text{There is some } y \in \Delta^{\mathcal{I}} \text{with } (x,y) \in r^{\mathcal{I}} \text{and } y \in C^{\mathcal{I}}\}$$

An interpretation satisfies a concept inclusion $C \sqsubseteq D$, if $C^{\mathcal{I}} \subseteq D^{\mathcal{I}}$, and $\mathcal{I}$ is a model of a TBox $\mathcal{T}$(denoted by $\mathcal{I} \models \mathcal{T}$) if $\mathcal{I}$ satisfies all concept inclusions in $\mathcal{T}$. A concept C is satisfiable relative to a TBox $\mathcal{T}$ if there is a model $\mathcal{I}$ of $\mathcal{T}$ such that $C^{\mathcal{I}} \neq \emptyset$. A concept C is subsumed by a concept D relative to a TBox $\mathcal{T}$(denoted by $\mathcal{T} \models C \sqsubseteq D$) if every model $\mathcal{I}$ of $\mathcal{T}$ satisfies the concept inclusion $C \sqsubseteq D$.

Let $\mathcal{EL}^{\neg}$ be the extension of $\mathcal{EL}$ by restricting the applicability of negation to concept names(atomic negation). More precisely, the $\mathcal{EL}^{\neg}$ concepts are built according to the following syntax rule:

$$C :: = \top | A | \neg A | C \sqcap D | \exists r.C,$$

where A ranges over concept names and r ranges over role names. An $\mathcal{EL}^{\neg}-$TBox is a finite set of concept inclusions $C \sqsubseteq D$, where C and D are $\mathcal{EL}^{\neg}-$concepts. Obviously, the description logic $\mathcal{EL}^{\neg}$ is a fragment of $\mathcal{ALC}$. As an example, here is a simple $\mathcal{EL}^{\neg}-$TBox $\mathcal{T}_1$:

Example 1. $\mathcal{T}_1$:

$$
\begin{aligned}
Mother &\equiv Female \sqcap \exists haschild.Person \\
Father &\equiv \neg Female \sqcap \exists haschild.Person \\
\neg Female &\sqsubseteq Person \\
Female &\sqsubseteq Person
\end{aligned}
$$

We introduce conservative extensions and inseparability of TBoxes. The notation of a signature, which is a finite subset of $\mathbf{N_C} \cup \mathbf{N_R}$. The signature $sig(C)$ of a concept C is the set of concept and role names occurring in C, and likewise for the signature $sig(\mathcal{T})$ of a TBox $\mathcal{T}$. Given a signature Σ, we use $\mathcal{C}(\Sigma)$ to denote the set of concepts using only concept and role names from Σ.

Definition 1. (Σ–**conservative extension**) *Let $\mathcal{T}_1$ and $\mathcal{T}_2$ be TBoxes with $\mathcal{T}_1 \subseteq \mathcal{T}_2$ and let $\Sigma \subseteq sig(\mathcal{T}_1)$. Then $\mathcal{T}_2$ is a conservative extension of $\mathcal{T}_1$ w.r.t. Σ if, for all $C, D \in \mathcal{C}(\Sigma)$, we have $\mathcal{T}_1 \models C \sqsubseteq D$, if and only if $\mathcal{T}_2 \models C \sqsubseteq D$.*

Obviously, by Definition 1, $\mathcal{T}_2$ is not a conservative extension of $\mathcal{T}_1$ if and only if there is a witness concept inclusion $C \sqsubseteq D$ such that

$$\mathcal{T}_1 \not\models C \sqsubseteq D, \text{but } \mathcal{T}_2 \models C \sqsubseteq D.$$

To illustrate the use of conservative extensions for controlling the consequences of ontology refinement, we give a simple example based on the description logic $\mathcal{EL}^{\neg}$. Let $\mathcal{T}_3$ be $\mathcal{T}_1$ extended with the following concept inclusions:

$$\exists haschild.\neg Female \sqsubseteq Parent$$
$$Parent \sqsubseteq Person$$

It is not difficult to check that $\mathcal{T}_3$ is not a conservative extension of $\mathcal{T}_1$ in $\mathcal{EL}^{\neg}$, as is shown by the witness concept inclusion $\exists haschild.\neg Female \sqsubseteq Person$. In other words, $\mathcal{T}_1 \not\models \exists haschild.\neg Female \sqsubseteq Person$, but $\mathcal{T}_3 \models \exists haschild.\neg Female \sqsubseteq Person$.

Definition 2. (Σ–**entailment, Σ–inseparability**) *Let $\mathcal{T}_1$ and $\mathcal{T}_2$ be TBoxes and Σ a signature. Then $\mathcal{T}_1$ Σ–entails $\mathcal{T}_2$, if $\mathcal{T}_2 \models C \sqsubseteq D$ implies $\mathcal{T}_2 \models C \sqsubseteq D$ for all concept inclusions $C \sqsubseteq D$ with $C, D \in \mathcal{C}(\Sigma)$. $\mathcal{T}_1$ and $\mathcal{T}_2$ are called Σ–inseparable if $\mathcal{T}_1$ and $\mathcal{T}_2$ Σ–entail each other.*

Note that if $\mathcal{T}_1 \subseteq \mathcal{T}_2$, then $\mathcal{T}_1$ Σ–entails $\mathcal{T}_2$. Therefore, the notion of Σ–concept entailment between TBoxes is a generalization of the notion of a conservative extension.

3   The Translation from $\mathcal{ALC}$ to $\mathcal{EL}^{\neg}$

The translation σ from $\mathcal{ALC}$ to $\mathcal{EL}^{\neg}$ consists two parts: for every concept C in $\mathcal{ALC}$, $\sigma(C)$ is a concept in $\mathcal{EL}^{\neg}$, and for every TBox $\mathcal{T}$ in $\mathcal{ALC}$, $\sigma(\mathcal{T})$ is a TBox in $\mathcal{EL}^{\neg}$. More precisely, the translation σ is defined as follows:

- For every $\mathcal{ALC}$ concept C :

$$\sigma(\top) = \top \qquad\qquad \sigma(A) = A, A \in N_C$$
$$\sigma(C \sqcap D) = \sigma(C) \sqcap \sigma(D) \qquad \sigma(\exists r.C) = \exists r.\sigma(C)$$
$$\sigma(\neg C) = \neg A' \qquad \text{where } A' \text{ is a new concept name}$$

- For every $\mathcal{ALC}$ TBox $\mathcal{T}$:

$$\sigma(C) \sqsubseteq \sigma(D) \in \sigma(\mathcal{T}) \text{ whenever } C \sqsubseteq D \in \mathcal{T}$$

  For $\neg C \in subcon(\mathcal{T})$ with C complex, we add the two concept inclusions $A' \sqsubseteq C$ and $C \sqsubseteq A'$ into $\sigma(\mathcal{T})$

Now, on the semantic side, for every interpretation $\mathcal{I}$ of $\mathcal{ALC}$, the corresponding interpretation $\mathcal{J}$ is defined as follows:

- $\Delta^{\mathcal{J}} = \Delta^{\mathcal{I}}$;
- for each concept name A, $A^{\mathcal{J}} = A^{\mathcal{I}}$;
- for each role name r, $r^{\mathcal{J}} = r^{\mathcal{I}}$;
- for each new concept name $A'^{\mathcal{J}} = C^{\mathcal{I}}$, where C is a complex concept with negation occurs.

For example, let $\mathcal{T}$ be a $\mathcal{ALC}$−TBox defined as follows:

Example 2. $\mathcal{T}$:

$$Mother \equiv Female \sqcap \exists haschild.Person$$
$$Father \equiv Male \sqcap \exists haschild.Person$$
$$Male \sqsubseteq Person$$
$$Female \sqsubseteq Person$$
$$Motherwithoutdaughter \sqsubseteq Mother \sqcap \forall haschild.\neg Male$$

Then the corresponding $\mathcal{EL}^{\neg}-$ TBox $\sigma(\mathcal{T})$ is defined as follows:

$$Mother \equiv Female \sqcap \exists haschild.Person$$
$$Father \equiv Male \sqcap \exists haschild.Person$$
$$Male \sqsubseteq Person$$
$$Female \sqsubseteq Person$$
$$Motherwithoutdaughter \sqsubseteq Mother \sqcap \neg A'$$
$$A' \sqsubseteq \exists haschild.Male$$
$$\exists haschild.Male \sqsubseteq A'$$

Compare $\mathcal{T}$ and $\sigma(\mathcal{T})$, we see that the translation σ gives rise to the signature symbols extension in $\mathcal{T}$, that is, fresh signature symbols are allowed to be introduced. In such case, the expressivity of a language can be increased.

We also note that the translation σ defined here is a refinement of the translation defined in Baader et al. [2] to prove the complexity of satisfiability and subsumption w.r.t general TBoxes. It is not hard to verify that this transformation can be done in polynomial time, yielding a $\mathcal{EL}^{\neg}$−TBox $\sigma(\mathcal{T})$ whose size is linear in the size of $\mathcal{T}$, where the size $|\mathcal{T}|$ of a $\mathcal{ALC}$−TBox is the number of symbols needed to write down $\mathcal{T}$. Using this translation, we get the following propositions:

Proposition 1. *Let* $\mathcal{T}$ *be a TBox and* C *a concept in* $\mathcal{ALC}$ *with* $sig(C) \subseteq sig(\mathcal{T})$. *Then* C *is satisfiable w.r.t.* $\mathcal{T}$ *in* $\mathcal{ALC}$, *if and only if* $\sigma(C)$ *is satisfiable w.r.t.* $\sigma(\mathcal{T})$ *in* $\mathcal{EL}^{\neg}$.

Proof. The proof is straightforward by induction on the structure of C. We only consider the interesting case of the induction, i.e., $C = \neg E$ with E complex.

- Let $\mathcal{I}$ be a model of $\mathcal{T}$ such that $C^{\mathcal{I}} \neq \emptyset$. Then we define a model $\mathcal{J}$ of $\sigma(\mathcal{T})$ by setting $\Delta^{\mathcal{J}} = \Delta^{\mathcal{I}}, A'^{\mathcal{J}} = E^{\mathcal{I}}$ where A' is a new concept name. For the other unnegated occurrences of concept names A and role names r in $\mathcal{T}$, we set $A^{\mathcal{J}} = A^{\mathcal{I}}, r^{\mathcal{J}} = r^{\mathcal{I}}$. Clearly, $\mathcal{J}$ is a model of $\sigma(\mathcal{T})$ and $\sigma(C)^{\mathcal{J}} \neq \emptyset$.

– Let $\mathcal{J}$ be a model of $\sigma(\mathcal{T})$ such that $\sigma(C)^{\mathcal{J}} \neq \emptyset$. Then we define a model $\mathcal{I}$ of $\mathcal{T}$ by setting $\Delta^{\mathcal{I}} = \Delta^{\mathcal{J}}, A^{\mathcal{I}} = A^{\mathcal{J}}, r^{\mathcal{I}} = r^{\mathcal{J}}$. Since $A' \sqsubseteq E$ and $E \sqsubseteq A'$, we have that $E^{\mathcal{I}} = A'^{\mathcal{J}}$. Hence, $I \models \mathcal{T}$ and $C^{\mathcal{I}} \neq \emptyset$.

By Proposition 1, we have that the satisfiability of concepts is preserved. However, translating in a satisfiability-preserving way does not immediately lead to the preservation of the unsatisfiability, if the translation between models is taken into account and the class of models of a source logical system is translated to a proper subclass of the class of models of a target logical system. For example, Fara and Williamson [4] showed that the translations from first-order modal logic into the counterpart theory given by Lewis [11], Forbes [6] and Ramachandran [22] may translate an unsatisfiable formula to a satisfiable formula, see [20] for more details.

Inspired by the observation, in [19,21], we give the following logical properties to describe a translation σ.

◇ *Faithfulness*: for every formula φ of a source logical system, φ is satisfied in a model $\mathcal{M}$ and a valuation v of the source logical system if and only if $\sigma(\varphi)$ is satisfied in the translated model $\sigma(\mathcal{M})$ and valuation $\sigma(v)$, that is,

$$(\mathcal{M}, v) \models \varphi \text{ if and only if } (\sigma(\mathcal{M}), \sigma(v)) \models \sigma(\varphi).$$

◇ *Fullness*: for any formula φ of a source logical system, any model $\mathcal{M}'$ and any valuation v' of a target logical system, if $(\mathcal{M}', v') \models \sigma(\varphi)$, then there exists a model $\mathcal{M}$ and a valuation v of the source logical system such that $(\mathcal{M}, v) \models \varphi$ and $\sigma(\mathcal{M}) = \mathcal{M}', \sigma(v) = v'$.

The faithfulness corresponds to the preservation of satisfiability. The fullness says that every model and every valuation which satisfies $\sigma(\varphi)$ has a corresponding model and valuation in the source logical system. By the definitions of faithfulness and fullness, we have that the satisfiability and unsatisfiability of formulas are both preserved.

Proposition 2. *Let $\mathcal{T}$ be a TBox and C a concept in $\mathcal{ALC}$ with $sig(C) \subseteq sig(\mathcal{T})$. Then C is unsatisfiable w.r.t. $\mathcal{T}$ in $\mathcal{ALC}$, if and only if $\sigma(C)$ is unsatisfiable w.r.t. $\sigma(\mathcal{T})$ in $\mathcal{EL}^{\neg}$.*

Proof. The proof is straightforward by induction on the structure of C. We only consider the interesting case of the induction, i.e., $C = \neg E$ with E complex.

– Suppose that C is unsatisfiable w.r.t. $\mathcal{T}$ in $\mathcal{ALC}$. If $\sigma(C)$ is satisfiable w.r.t. $\sigma(\mathcal{T})$, then there exists a model $\mathcal{J}$ of $\sigma(\mathcal{T})$ such that $\sigma(C)^{\mathcal{J}} = (\neg A')^{\mathcal{J}} \neq \emptyset$. We define a model $\mathcal{I}$ of $\mathcal{T}$ by setting $\Delta^{\mathcal{I}} = \Delta^{\mathcal{J}}, A^{\mathcal{I}} = A^{\mathcal{J}}, r^{\mathcal{I}} = r^{\mathcal{J}}$. Since $A' \sqsubseteq E$ and $E \sqsubseteq A'$, we have that $E^{\mathcal{I}} = A'^{\mathcal{J}}$. Hence, $I \models \mathcal{T}$ and $C^{\mathcal{I}} \neq \emptyset$.
– Suppose that $\sigma(C)$ is unsatisfiable w.r.t. $\sigma(\mathcal{T})$ in $\mathcal{EL}^{\neg}$. If C is satisfiable w.r.t. $\mathcal{T}$, then there exists a model $\mathcal{I}$ of $\mathcal{T}$ such that $C^{\mathcal{I}} \neq \emptyset$. We define $\mathcal{J}$ of $\sigma(\mathcal{T})$ by setting $\Delta^{\mathcal{J}} = \Delta^{\mathcal{I}}, A'^{\mathcal{J}} = E^{\mathcal{I}}$ where A' is a new concept name. For the other unnegated occurrences of concept names A and role names r in $\sigma(\mathcal{T})$, we set $A^{\mathcal{J}} = A^{\mathcal{I}}, r^{\mathcal{J}} = r^{\mathcal{I}}$. Clearly, $\mathcal{J}$ is a model of $\sigma(\mathcal{T})$ and $\sigma(C)^{\mathcal{J}} \neq \emptyset$.

Propositions 1 and 2 show that the translation σ defined in the section is a faithful and full translation from $\mathcal{ALC}$ to $\mathcal{EL}^{\neg}$.

Proposition 3. *Let* $\mathcal{T}$ *be a TBox and* C, D *concepts in* $\mathcal{ALC}$ *with* $sig(C), sig(D)$ $\subseteq sig(\mathcal{T})$. *Then* $\mathcal{T} \models C \sqsubseteq D$, *if and only if* $\sigma(\mathcal{T}) \models \sigma(C) \sqsubseteq \sigma(D)$.

Proof. We only consider the interesting case of $\neg E$ with complex E, which can be replaced with $\neg A'$ for a new concept name A' if we add two concept inclusions $A' \sqsubseteq E$ and $E \sqsubseteq A'$.

- Suppose that $\mathcal{T} \models C \sqsubseteq D$ with $sig(C), sig(D) \subseteq sig(\mathcal{T})$. For every model $\mathcal{J}$ of $\sigma(\mathcal{T})$, we need to show $\mathcal{J} \models \sigma(C) \sqsubseteq \sigma(D)$. If it is not, i.e., there is a model $\mathcal{J}$ such that $\mathcal{J} \models \sigma(\mathcal{T})$ but $\sigma(C)^{\mathcal{J}} \not\subseteq \sigma(D)^{\mathcal{J}}$. Then we define a model $\mathcal{I}$ by setting $\Delta^{\mathcal{I}} = \Delta^{\mathcal{J}}, A^{\mathcal{I}} = A^{\mathcal{J}}, r^{\mathcal{I}} = r^{\mathcal{J}}$. Since $A' \sqsubseteq E$ and $E \sqsubseteq A'$, we set $E^{\mathcal{I}} = A'^{\mathcal{J}}$. Hence $\mathcal{I} \models \mathcal{T}$, but $\mathcal{I} \not\models C \sqsubseteq D$.
- Suppose that $\sigma(\mathcal{T}) \models \sigma(C) \sqsubseteq \sigma(D)$, but $\mathcal{T} \not\models C \sqsubseteq D$. Then there exists a model $\mathcal{I}$ such that $\mathcal{I} \models \mathcal{T}$ but $C^{\mathcal{I}} \not\subseteq D^{\mathcal{I}}$. We define a model of $\mathcal{J}$ of $\sigma(\mathcal{T})$ by setting $\Delta^{\mathcal{J}} = \Delta^{\mathcal{I}}, A'^{\mathcal{J}} = E^{\mathcal{I}}$, where A' is a new concept name. For the other unnegated occurrences of concept names A and role names r in $\mathcal{T}$, we set $A^{\mathcal{J}} = A^{\mathcal{I}}, r^{\mathcal{J}} = r^{\mathcal{I}}$. Clearly, $\mathcal{J} \models \sigma(\mathcal{T})$ but $\mathcal{J} \not\models \sigma(C) \sqsubseteq \sigma(D)$.

4   Complexity Results

In this section, we reduce the problem of deciding conservative extensions and inseparability in $\mathcal{ALC}$ can be reduced to the case of $\mathcal{EL}^{\neg}$, and these results establish that 2ExpTime-completeness of both inseparability and conservative extensions in $\mathcal{EL}^{\neg}$.

The upper bound follows from $\mathcal{EL}^{\neg}$ being a fragment of $\mathcal{ALC}$. For the lower bound, we reduce conservative extensions of $\mathcal{ALC}$ w.r.t. general TBoxes to the case of $\mathcal{EL}^{\neg}$.

Theorem 1. *Let* $\mathcal{T}_1$ *and* $\mathcal{T}_2$ *be two TBoxes in* $\mathcal{ALC}$ *with* $\mathcal{T}_1 \subseteq \mathcal{T}_2$. *Then* $\mathcal{T}_2$ *is a conservative extension of* $\mathcal{T}_1$ *w.r.t.* Σ, *if and only if* $\sigma(\mathcal{T}_2)$ *is a conservative extension of* $\sigma(\mathcal{T}_1)$ *w.r.t.* $\sigma(\Sigma)$, *where* $\sigma(\Sigma) = \Sigma \cup \{ A' \mid$ *where* A' *is a new concept name*$\}$.

Proof. Suppose that $\mathcal{T}_2$ is not a conservative extension of $\mathcal{T}_1$ w.r.t. Σ, there exists a concept inclusion $C \sqsubseteq D$ with $C, D \in \mathcal{C}(\Sigma)$ such that $\mathcal{T}_1 \not\models C \sqsubseteq D$ but $\mathcal{T}_2 \models C \sqsubseteq D$. By Proposition 3, we have that $\sigma(\mathcal{T}_1) \not\models \sigma(C) \sqsubseteq \sigma(D)$ and $\sigma(\mathcal{T}_2) \models \sigma(C) \sqsubseteq \sigma(D)$. That is, $\sigma(\mathcal{T}_2)$ is not a conservative extension of $\sigma(\mathcal{T}_1)$ w.r.t $\sigma(\Sigma)$.

Suppose that $\sigma(\mathcal{T}_2)$ is not a conservative extension of $\sigma(\mathcal{T}_1)$ w.r.t. $\sigma(\Sigma)$, there exists a concept inclusion $C' \sqsubseteq D'$ with $C', D' \in \mathcal{C}(\sigma(\Sigma))$ such that $\sigma(\mathcal{T}_1) \not\models C' \sqsubseteq D'$ but $\sigma(\mathcal{T}_2) \models C' \sqsubseteq D'$. Next, we consider the two cases of the concept inclusion $C' \sqsubseteq D'$:

- Assume that the new concept name $A' \notin sig(C') \cup sig(D')$. We have to show that $\mathcal{T}_1 \not\models C' \sqsubseteq D'$ but $\mathcal{T}_2 \models C' \sqsubseteq D'$. We note that C' and D' are $\mathcal{EL}^{\neg}$ concepts, by the translation σ, $\sigma(C') = C', \sigma(D') = D'$. Therefore, by Proposition 3, $\mathcal{T}_1 \not\models C' \sqsubseteq D', \mathcal{T}_2 \models C' \sqsubseteq D'$.

- Suppose that the new concept name $A' \in sig(C') \cup sig(D')$. For each new concept name A' occurs in $C' \sqsubseteq D'$, we replace A' with the complex concept E, which satisfies $A' \sqsubseteq E, E \sqsubseteq A'$. Then, by the translation σ, $T_1 \not\models C'_E \sqsubseteq D'_E$ but $T_2 \models C'_E \sqsubseteq D'_E$, where C'_E and D'_E are the replacement concepts of C' and D', respectively.

By Theorem 1, we get the following complexity result of conservative extensions in $\mathcal{EL}^\neg$:

Theorem 2. *In $\mathcal{EL}^\neg$, conservative extensions w.r.t. general TBoxes is 2ExpTime-complete.*

Next, we set up the complexity of the problem of inseparability. The lower bound of inseparability is 2ExpTime, since the notion of Σ–concept entailment between TBoxes is a generalization of the notion of conservative extension. The upper bound follows that the problem of inseparability of $\mathcal{ALC}$ is 2ExpTime-complete.

Theorem 3. *In $\mathcal{EL}^\neg$, the inseparability w.r.t. general TBoxes is 2ExpTime-complete.*

5 Conclusion and Future Work

In this paper, we investigate the complexity problem of deciding whether the extension of an ontology is conservative w.r.t. the signature Σ. A generalization case is to decide whether two ontologies are inseparable. By adding many countable new concept names which corresponds to the complex negative concepts, we establish a translation from $\mathcal{ALC}$ to $\mathcal{EL}^\neg$ and reduce the problem of conservative extensions in $\mathcal{ALC}$ to the case of $\mathcal{EL}^\neg$. Based on the result, we establish 2ExpTime-completeness of both conservative extensions and inseparability.

In future work we will consider the complexity problem of conservative extensions and inseparability in other extensions of $\mathcal{EL}$, that is, by adding role or other concept constructors.

Acknowledgments. The work was supported by the National Natural Science Foundation of China under Grant Nos.60573010, 61103169.

References

1. Baader, F., Nutt, W.: Basic description logics. In: Baader, F., Calvanese, D., McGuinness, D., Nardi, D., Patel-Scheider, P.F. (eds.) The Description Logic Handbook: Theory, Implementation, and Applications. Cambridge University Press, Cambridge (2003)
2. Baader, F., Brandt, S., Lutz, C.: Pushing the $\mathcal{EL}$ envelope. In: Proceedings of the 19th International Joint Conference on Artificial Intelligence (IJCAI'05), pp. 364–369. AAAI Press (2005)

3. Diaconescu, R., Goguen, J., Stefaneas, P.: Logical support for modularisation. In: Huet, G., Plotkin, G. (eds.) Logical Environments. Cambridge University Press, New York (1993)
4. Fara, M., Williamson, T.: Counterparts and actuality. Mind **114**, 1–30 (2005)
5. Fink, M.: Equivalences in answer-set programming by countermodels in the logic of here-and-there. In: Garcia de la Banda, M., Pontelli, E. (eds.) ICLP 2008. LNCS, vol. 5366, pp. 99–113. Springer, Heidelberg (2008)
6. Forbes, G.: Canonical counterpart theory. Analysis **42**, 33–37 (1982)
7. Ghilardi, S., Lutz, C., Wolter, F.: Did I damage my ontology? a case for conservative extensions in description logics. In: Proceedings of KR06, pp. 187–197. AAAI Press (2006)
8. Grau, B.C., Horrocks, I., Kazakov, Y., Sattler, U.: Modular reuse of ontologies: theory and practice. J. Artif. Intell. Res. **31**, 273–318 (2008)
9. Konev, B., Lutz, C., Walther, D., Wolter, F.: Semantic modularity and module extraction in description logics. In: Proceedings of ECAI'08, pp. 55–59 (2008)
10. Kontchakov, R., Wolter, F., Zakharyaschev, M.: Logic-based ontology comparison and module extraction with an application to DL-Lite. J. Artif. Intell. **174**, 1093–1141 (2010)
11. Lewis, D.: Counterpart theory and quantified modal logic. J. Philos. **65**, 113–126 (1968)
12. Lutz, C., Walther, D., Wolter, F.: Conservative extensions in expressive description logics. In: Proceedings of IJCAI07, pp. 453–458. AAAI Press (2007)
13. Lutz, C., Wolter, F.: Deciding inseparability and conservative extensions in the description logic EL. J. Symbolic Comput. **45**, 194–228 (2010)
14. Maibaum, T.: Conservative extensions, interpretations between theories and all that!. In: Bidoit, M., Dauchet, M. (eds.) CAAP, FASE and TAPSOFT 1997. LNCS, vol. 1214, pp. 40–66. Springer, Heidelberg (1997)
15. Pearce, D., Valverde, A.: Synonymous theories in answer set programming and equilibrium logic. In: Proceedings of the 16th European Conference on Artificial Intelligence (ECAI 2004), pp. 388–392 (2004)
16. Sioutos, N., de Coronado, S., Haber, M., Hartel, F., Shaiu, W., Wright, L.: NCI thesaurus: a semantic model integrating cancer-related clinical and molecular information. J. Biomed. Inform. **40**, 30–43 (2006)
17. Spackman K. Managing clinical terminology hierarchies using algorithmic calculation of subsumption: Experience with SNOMED-RT. (2000) Fall Symposium Special Issue
18. Blake, J.A., et al.: The gene ontology: enhancements for 2011. Nucleic Acids Res. **40**, 559–564 (2012)
19. Shen, Y., Ma, Y., Cao, C., Sui, Y., Wang, J.: Logical properties on translations between logics. Chin. J. Comput. **32**, 2091–2098 (2009)
20. Shen, Y., Sui, Y., Wang, J.: On the translation from quantified modal logic into the counterpart theory revisited. In: Xiong, H., Lee, W.B. (eds.) KSEM 2011. LNCS, vol. 7091, pp. 377–386. Springer, Heidelberg (2011)
21. Shen, Y., Ma, Y., Cao, C., Sui, Y., Wang, J.: Faithful and full translations between logics. Ruan Jian Xue Bao/J. Softw. **24**, 1626–1637 (2013). (in Chinese)
22. Ramachandran, M.: An alternative translation scheme for counterpart theory. Analysis **49**, 131–141 (1989)

GrOD: Graph-based Ontology Debugging System

Xuefeng Fu$^{(\boxtimes)}$, Yong Zhang, and Guilin Qi

School of Computer Science and Engineering, Southeast University,
Nanjing 210096, China
{fxf,zhangyong,gqi}@seu.edu.cn

Abstract. In this paper, we present *GrOD*, a Graph-based Ontology Debugging System for DL-Lite ontologies, which implements a graph-based algorithm for ontology debugging. *GrOD* encodes ontology into directed graph and stores them in Neo4j graph database. It debugs incoherence of the ontology by finding whether there exist two paths that from common node to two disjoint nodes respectively on the graph. Our demonstration will illustrate functionalities of *GrOD* for computing MUPS (minimal unsatisfiablility-preserving subterminology) and MIPS (minimal incoherence-preserving subterminology).

1   Introduction

Ontology is a formal specification of a shared conceptualization, it allows information to be shared in a semantically unambiguous way and domain knowledge to be reused [8]. However, in the processing of ontology evolution, incorporating newly received information that from different sources would lead the ontology incoherent. Reasoning on the incoherent ontologies could infer arbitrary conclusion. Therefore, Ontology debugging in description logics has become one of the key task in the semantic web.

Recently, there has been some debugging work on lightweight family of DLs, called DL-Lite family, which is a family of DLs that provides tractable reasoning services [2]. Several ontology debugging methods [5,6,8] have been proposed. In general, the main target of these methods is to obtain the explanation of unsatisfiable concept(role) which is a subset of terminologies. They can be distinguished into black-box approaches, which use an external reasoner for testing whether certain combinations of axioms may cause incoherences, and glass-box approaches that modify the reasoning procedure in order to improve the information returned [5]. These approaches compute the unsatisfiable concepts(roles), and then infer all minimal unsatisfiability-preserving subterminology (MUPS) of unsatisfiable concepts(roles). From the MUPS, they can compute all minimal incoherence-preserving subterminology (MIPS) of ontology.

In this paper we describe an implementation of our graph-based algorithm and develop a system, called *GrOD*, which focuses on the debugging of terminological part of ontologies. We first introduce preliminary theory of DL-Lite family, and then demonstrate our implementation for ontology debugging.

© Springer-Verlag Berlin Heidelberg 2014
D. Zhao et al. (Eds.): CSWS 2014, CCIS 480, pp. 87–94, 2014.
DOI: 10.1007/978-3-662-45495-4_8

2   Graph-Based Terminology Debugging

2.1   DL-Lite

In our work, we consider DL-Lite$_{FR}$, which is an important component in DL-Lite family. We start with the introduction of DL-Lite$_{core}$[2], which is the core language for the DL-Lite family [1]. The complex concepts and roles of DL-Lite$_{core}$ are defined as follows: $(1)B \rightarrow A \mid \exists R$, $(2)R \rightarrow P \mid P^-$, $(3)C \rightarrow B \mid \neg B$, $(4)E \rightarrow R \mid \neg R$, where A denotes an atomic concept, P an atomic role, B a basic concept, C a general concept, R a basic role and E a general role. A basic concept which can be either an atomic concept or a concept of the form $\exists R$, where R denotes a basic role which can be either an atomic role or the inverse of an atomic role.

DL-Lite$_R$ extends DL-Lite$_{core}$ with inclusion assertions between roles of the form $R \sqsubseteq E$, where R and E are defined as above. DL-Lite$_F$ extends DL-Lite$_{core}$ with functional restriction on roles or on their inverses with the form $(functR)$ or $(functR^-)$.

The semantics of DL-Lite is defined by an interpretation $\mathcal{I} = (\triangle^{\mathcal{I}}, \cdot^{\mathcal{I}})$ which consists of a non-empty domain set $\triangle^{\mathcal{I}}$ and an interpretation function $\cdot^{\mathcal{I}}$, which maps from individuals, concepts and roles to elements of the domain, subsets of the domain and binary relations on the domain, respectively.

2.2   Graph Construction

Given a DL-Lite ontology $\mathcal{O}$ over a signature Σ, which can be partitioned into two disjoint signatures, Σ_T, containing symbols for atomic elements, i.e., atomic concept and atomic roles, and Σ_C, containing symbols for individuals. In this paper, we only focus on the terminological part of ontology.

According to the work given in [3], the digraph $G_T = \langle N, E \rangle$ constructed from terminological part of ontology over the signature Σ is given as follows:

(1) for each atomic concept B in Σ_T, N contains the node B.
(2) for each atomic role P in Σ_T, N contains the node $P, P^-, \exists P, \exists P^-$.
(3) for each concept inclusion $B_1 \sqsubseteq B_2 \in T$, E contains the arc (B_1, B_2).
(4) for each concept inclusion $B_1 \sqsubseteq \neg B_2 \in T$, E contains the arc (B_1, B_2) and N contains the node $\neg B_2$.
(5) for each role inclusion $R_1 \sqsubseteq R_2 \in T$, E contains the arc (R_1, R_2), arc (R_1^-, R_2^-), arc $(\exists R_1, \exists R_2)$, arc $(\exists R_1^-, \exists R_2^-)$.
(6) for each role inclusion $R_1 \sqsubseteq \neg R_2 \in T$, E contains the arc $(R_1, \neg R_2)$, arc $(R_1^-, \neg R_2^-)$, arc $(\exists R_1, \neg \exists R_2)$, arc $(\exists R_1^-, \neg \exists R_2^-)$ and N contains nodes $\neg R_2$, $\neg R_2^-$, $\neg \exists R_2$, $\neg \exists R_2^-$.

In items (1)–(6), we construct the graph based on TBox $\mathcal{T}$. It has been shown in [4] that the problem of TBox classification in a DL-Lite ontology $\mathcal{O}$ can be done by computing transitive closure of the graph $\mathcal{G}_T$. For simplicity, we call these rules **Construction Rules**, and we use node C (node R) to denote the node w.r.t. concept C (role R).

In our graph, each node represents a basic concept or a basic role, while each arc represents an inclusion assertion, i.e. the start node of the arc corresponds to the left-hand side of the inclusion assertion and the end node of the arc corresponds to the right-hand side of the inclusion assertion. In order to ensure that the information represented in the TBox is preserved by the graph, we add nodes R, R^-, $\exists R$, $\exists R^-$ for each role R, $arc(R_1, R_2)$, $arc(R_1^-, R_2^-)$, $arc(\exists R_1, \exists R_2)$, $arc(\exists R_1^-, \exists R_2^-)$ for each role inclusion assertion $R_1 \sqsubseteq R_2$.

2.3 Equivalence of Ontology and Graph

From the rule of constructing graph, the path in the graph is corresponding to a subset of TBox and the reasoning of DL-Lite ontology is logical equivalence to reachability between nodes of graph. Let $\mathcal{T}$ be an DL-Lite$_{\mathcal{FR}}$ TBox and let $G_{\mathcal{T}} = \langle N, E \rangle$ be the digraph constructed from $\mathcal{T}$ according to Construction Rules, we give following property by extending work in [4].

Theorem 1. *Let m and n be two basic concepts, or two basic roles. $m \sqsubseteq n \in cl(\mathcal{T})$ if and only if $arc(m, n) \in E^*$*

From above assumption, we give corresponding definition of MUPS and MIPS on the graph.

Definition 1. *Let $\mathcal{T}$ be an DL-Lite$_{\mathcal{FR}}$ TBox and let $G_{\mathcal{T}} = \langle N, E \rangle$ be the digraph constructed from $\mathcal{T}$ according to Construction Rules, i.e., each node represents a concept or role and each edge represents an inclusion assertion. Then for arbitrary node $C \in N$, if there exist two paths that are from node C to node D and from node C to node $\neg D$ respectively in $G_{\mathcal{T}}$, we call these two paths minimal unsatisfiability-preserving path-pair (MUPP) w.r.t. node C.*

We use $mupp(C, \mathcal{T})$ to denote the set of all MUPP w.r.t node C in $G_{\mathcal{T}}$.

Definition 2. *Let $\mathcal{T}$ be an DL-Lite$_{\mathcal{FR}}$ TBox and let $G_{\mathcal{T}} = \langle N, E \rangle$ be the digraph constructed from $\mathcal{T}$ according to Construction Rules. Then for arbitrary node $C \in N$, if there exist two paths that are from node C to node D and from node C to node $\neg D$ respectively in $\mathcal{G}_{\mathcal{T}}$ and there does not exist joint edge in $C \rightarrow D$ and $C \rightarrow \neg D$, we call these two paths as minimal incoherence-preserving path-pair (MIPP).*

Obviously, a MIPP is definitely a MUPP. We use mipp to denote the set of all MIPP in $G_{\mathcal{T}}$. MUPP and MIPP holds the following properties.

Theorem 2. *Let $\mathcal{S}'$ be the subset of $\mathcal{T}$ that corresponds to a MUPP w.r.t. node C in $\mathcal{T}$. Then $\mathcal{S}'$ is a MUPS w.r.t. C in $\mathcal{T}$.*

Theorem 3. *Let $\mathcal{S}'$ be the subset of $\mathcal{T}$ that corresponds to a MIPP in $\mathcal{T}$. Then $\mathcal{S}'$ is a MIPS in $\mathcal{T}$.*

The existence of MUPP(or MIPP) indicates that the ontology is incoherent.

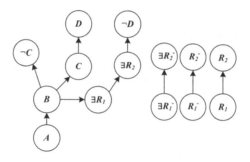

Fig. 1. Directed graph in example 1

Theorem 4. *Let $\mathcal{T}$ be a DL-Lite$_{\mathcal{FR}}$ TBox, $\mathcal{T}$ is incoherent iff there exists at least one MUPP (or MIPP) in $\mathcal{G}_{\mathcal{T}}$.*

Obviously, above theorems indicate that compute MUPS and MIPS of ontology can be shift to compute MUPP and MIPP on the graph.

In the following, we can compute MUPS and MIPS on the graph. Let S be a concept or role expression. At first, we find all descendant nodes of node S and node $\neg S$ in graph $\mathcal{G}_{\mathcal{T}}$. If there is a common node, which we assume as S', between descendant nodes of node S (Specially, it includes S itself for the case $S \sqsubseteq \neg S$) and those of node $\neg S$, the common node S' is a unsatisfiable node. There two paths $S' \to S$ and $S' \to \neg S$ in $\mathcal{G}_{\mathcal{T}}$ is MUPP of S'. If $S' \to S$ and $S' \to \neg S$ have no joint edge, then $\{S' \to S, S' \to \neg S\}$ is a MIPP.

We illuminate the process of computing the MUPP and MIPP as follow.

Example 1. Given a TBox $\mathcal{T}$, where $\mathcal{T} = \{A \sqsubseteq B, B \sqsubseteq C, C \sqsubseteq D, B \sqsubseteq \neg C, R_1 \sqsubseteq R_2, B \sqsubseteq \exists R_1, \exists R_2 \sqsubseteq \neg D\}$. According to the Construction Rules, we construct a directed graph shown in Fig. 1.

In Fig. 1, there have two disjointness pairs of node $(C, \neg C)$, $(D, \neg D)$. we can get from the graph that node B and node A are unsatisfiable nodes because there are descendant nodes of disjointness pairs. Then we can find two MIPPs in Fig. 1:$\{B \to \neg C, B \to C\}$ and $\{B \to D, B \to \neg D\}$ and two MUPPs in Fig. 1: $\{A \to \neg C, A \to C\}$ and $\{A \to D, A \to \neg D\}$. Since $\exists R_1 \sqsubseteq \exists R_2 \notin \mathcal{T}$ and $R_1 \sqsubseteq R_2 \models \exists R_1 \sqsubseteq \exists R_2$, $arc(\exists R_1, \exists R_2)$ corresponds to $R_1 \sqsubseteq R_2$ in $\mathcal{T}$. Finally, two MIPSs $\{B \sqsubseteq C, B \sqsubseteq \neg C\}$ and $\{B \sqsubseteq C, C \sqsubseteq D, R_1 \sqsubseteq R_2, B \sqsubseteq \exists R_1, \exists R_2 \sqsubseteq \neg D\}$ and two MUPSs $\{A \sqsubseteq B, B \sqsubseteq C, B \sqsubseteq \neg C\}$ and $\{A \sqsubseteq B, B \sqsubseteq C, C \sqsubseteq D, R_1 \sqsubseteq R_2, B \sqsubseteq \exists R_1, \exists R_2 \sqsubseteq \neg D\}$ will be generated.

3 Experimental Evaluation

We evaluate the performance of *GrOD* over adapted real ontologies. We compare *GrOD* with the systems that implementing the method for computing MIPS given in [8] by using different reasoners, such as Pellet and FaCT++. All experiments were performed on a *Lenovo* desktop PC with Intel Corei5-2400 3.1 GHz

Table 1. Experimental ontologies

Ontology	Axioms	Unsatisfiable Concepts	MIPS
Economy	803	51	47
Terrorism	185	14	5
Transportation	1186	62	36
Aeo	521	49	17
CL	8783	59	25
DOLCE-LITE	98	33	5
Fly-Anatomy	17735	304	3
FMA	160936	45	12
GO	43934	97	18
Plant	3295	45	12

CPU and 4 GB of RAM, running Microsoft window 7 operating system, and Java 1.7 with 3 GB of heap space.

Ontologies used in our experiments have significantly different sizes and structures. In order to fit $DL\text{-}Lite_{FR}$ expressivity, when an ontology cannot be expressed by $DL\text{-}Lite_{FR}$, it will be approximated [4]. Furthermore, since these ontologies are coherent, we modified them by inserting some "incoherent-generating" axioms randomly, such as disjointness axioms of the form $A \sqsubseteq \neg A'$. Table 1 lists statistics of some typical ontologies used in the experiments.

Table 2 shows the detailed performance comparison on different ontologies. We carried out these experiments on ten ontologies. According to the result, our system can compute MIPS of the test ontologies very efficiently. Furthermore, GrOD outperforms systems implementing the state of the art debugging algorithm. In particular, it performs one order of magnitude faster than existing systems for complex ontology FMA.

4   Implementation and Demonstration

We have implemented our graph-based algorithm for TBox debugging in the Graph-based Ontology Debugging System (*GrOD*) in Java. Ontology will be encoded to a directed graph and stored in Neo4j [9]. Neo4j is an open-source and high-performance graph database supported by Neo Technology. The main ingredients of Neo4j are nodes and relationships. A graph database is used to record data in nodes which have user-defined properties, while nodes are organized by relationships which also have user-defined properties. Users could look up nodes or relationships by index. Cypher is a powerful declarative graph query language [7]. We can leverage its ability to expressive and efficient querying and updating of the graph store. In our implementation, we apply Cypher to find the path between disjointness nodes.

Table 2. Time required to find all MIPS in milliseconds on different ontologies

Ontology	Pellet	Hermit	FaCT++	Jfact	Graph
Economy	1286	1378	619	1836	601
Terrorism	388	449	387	455	157
Transportation	1772	2735	1339	4712	618
Aeo	1287	1963	792	1517	357
CL	1465	2375	1152	2269	628
DOLCE-Lite	765	1041	475	813	325
Fly-Anatomy	2019	3319	1878	4901	590
FMA	47262	49444	45559	91776	3674
GO	3765	5371	3796	9315	875
Plant	1543	2009	935	1883	513

GrOD provides interface to get the incoherent information and to calculate explanation of unsatisfiable concept(role) in ontology. As shown in left of Fig. 1, *GrOD* has two major modules: (1) ontology management; (2) ontology debugging. In the module of "Ontology", users can upload ontology files to the system, UI of "Ontology" is shown in Fig. 2.

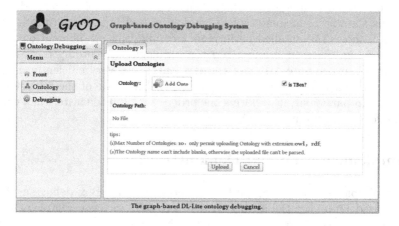

Fig. 2. The UI of ontology uploading

In the module of "Debugging", *GrOD* provides four functionalities:

(1) **ShowTree.** In this part, hierarchical information of ontology will be demonstrate into a tree structure (see right part of Fig. 3).
(2) **Transform.** In this part, the selected ontology will be encoded into graph structure and be store in neo4j database in the background.

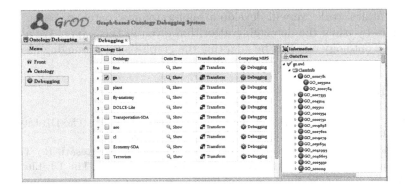

Fig. 3. The UI of debugging

(3) **Debugging.** *GrOD* can carry out the debugging of TBox and the result of debugging will be returned from service.

(4) **Explanation.** *GrOD* also provides the explanation which is a subset of TBox to explain why the ontology is incoherent. User can select number of MIPS to get the detail of explain, as shown in Fig. 4.

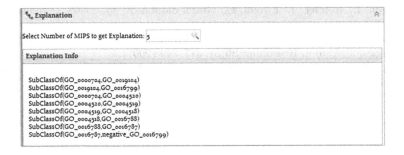

Fig. 4. The UI of detail of explain

5 Conclusion

We developed a graph-based system, *GrOD*, for DL-Lite ontologies debugging. It is able to calculate MUPS and MIPS of inconherent ontology. In the system, we translate concepts(roles) into vertices of graph and axiom into arcs of graph. *GrOD* is coded by java and neo4j graph database, which can be deployed on the web and can be convenient accessed by browser. We evaluated the system and the result showed that our method has gained improvement over several well-known tableau-based algorithms both in efficiency and stability.

As a future work, we will develop a semi-automatic ontology repair system in DL-Lite by using the debugging algorithm given in this paper. We also plan to work on ontology merging by adapting the graph-based debugging algorithm.

References

1. Artale, A., Calvanese, D., Kontchakov, R., Zakharyaschev, M.: The DL-Lite family and relations. J. Artif. Intell. Res. (JAIR) **36**, 1–69 (2009)
2. Calvanese, D., De Giacomo, G., Lembo, D., Lenzerini, M., Rosati, R.: Tractable reasoning and efficient query answering in description logics: The DL-Lite family. J. Autom. Reasoning **39**(3), 385–429 (2007)
3. Gao, S., Qi, G., Wang, H.: A new operator for abox revision in DL-Lite. In: Proceedings of AAAI'12, pp. 2423–2424 (2012)
4. Lembo, D., Santarelli, V., Savo, D.F.: Graph-based ontology classification in OWL 2 QL. In: Cimiano, P., Corcho, O., Presutti, V., Hollink, L., Rudolph, S. (eds.) ESWC 2013. LNCS, vol. 7882, pp. 320–334. Springer, Heidelberg (2013)
5. Parsia, B., Sirin, E., Kalyanpur, A.: Debugging OWL ontologies. In: Proceedings of WWW'05, pp. 633–640. ACM (2005)
6. Qi, G., Hunter, A.: Measuring incoherence in description logic-based ontologies. In: Aberer, K., Choi, K.-S., Noy, N., Allemang, D., Lee, K.-I., Nixon, L.J.B., Golbeck, J., Mika, P., Maynard, D., Mizoguchi, R., Schreiber, G., Cudré-Mauroux, P. (eds.) ASWC 2007 and ISWC 2007. LNCS, vol. 4825, pp. 381–394. Springer, Heidelberg (2007)
7. Robinson, I., Webber, J., Eifrem, E.: Graph databases, pp. 25–63. O'Reilly Media Inc.,(2013)
8. Schlobach, S., Huang, Z., Cornet, R., Van Harmelen, F.: Debugging incoherent terminologies. J. Autom. Reasoning **39**(3), 317–349 (2007)
9. Webber, J.: A programmatic introduction to neo4j. In Proceedings of SPLASH'12, pp. 217–218 (2012)

Semantic Data Generation
and Management

Effective Chinese Organization Name Linking to a List-Like Knowledge Base

Chengyuan Xue[1], Haofen Wang[1(✉)], Bo Jin[2], Mengjie Wang[1], and Daqi Gao[1]

[1] East China University of Science and Technology, Shanghai 200237, China
{030120725,whfcarter,10101507,
gaodaqi}@mail.ecust.edu.cn
[2] The 3rd Research Institute of the Public Security Ministry,
Shanghai 200031, China
jinbo@stars.org.cn

Abstract. Entity Linking is widely used in entity retrieval and semantic search. It refers mentions in unstructured documents to their representations in a knowledge base (KB). The frequently used KB (e.g. Wikipedia) usually contains abundant information corresponding to each entity, such as properties, name variations and text descriptions, which can help to find candidates and disambiguate the links. In this paper, we link organization names in Chinese documents to a list-like KB. Compared to typical KBs, the records in our KB are simply Chinese organization full names. The massive variations, or abbreviations in the documents cannot be directly matched to any organization name in the KB and bring about ambiguities, thus make the linking task difficult. At first, we enrich the KB with the abbreviations. Making use of the information from Hudong Baike and other sources, we design a pattern based full name annotation method to help generate abbreviations for all the names in the KB. To resolve the ambiguity problem, we propose a two-stage linking generation approach utilizing the co-occurrence of abbreviations and full names in the same document or document cluster, where the linked full names in the first stage constraint the linking of abbreviations in the second stage. We apply our approach to police inquiry document corpus. The experiment results show the effectiveness of our approach and outperforms the one-stage approach significantly in terms of precision and recall.

1 Introduction

Given a plain text, the Entity Linking (EL) task aims to identify small fragments of a text (commonly known as mentions) referring to entities in a given knowledge base (KB), e.g. Wikipedia. EL has become a key enabler for many intelligent applications, such as entity retrieval, semantic search and ontology learning. A typical EL system accomplishes the task in three steps: (a) *mention detection*, that is, determining linkable phrases in the text; (b) *selecting and ranking candidate entities* in the KB that are referred by each mention; (c) *disambiguation*, selecting the most possible candidate to link. Wikipedia is a widely-used KB for EL. It contains plenty additional information for entities. The properties from infoboxes and text descriptions from abstracts in entity

© Springer-Verlag Berlin Heidelberg 2014
D. Zhao et al. (Eds.): CSWS 2014, CCIS 480, pp. 97–110, 2014.
DOI: 10.1007/978-3-662-45495-4_9

pages, as well as the corresponding redirect pages and the disambiguation pages, can all be used for describing these entities and be leveraged for EL.

In this paper, we try to solve a different EL problem in which we only have a list-like KB. More precisely, we refer mentions in a document collection to Chinese organization names (denoted as CONs) listed in a KB. Linking mentions to a list-like KB happens if CONs are exported from some yellow pages or dictionaries in which most CONs are not covered by widely used KBs like Wikipedia or need a lot of efforts to be linked to such KBs. Compared with rich KBs (e.g. Wikipedia) containing plenty of entity information, ours is just a list of organizations with their full names without any additional information, making our Entity Linking task not trivial.

There are two main challenges remaining in the new EL problem. Firstly, a CON may be written in various variations or abbreviations. For example, abbreviations like "安吉公司" (Anji Co.), "上海安吉" (Shanghai Anji), "安吉广告" (Anji Ad), and "安吉" (Anji) can all represent the organization "上海安吉广告有限公司" (Shanghai Anji Ad Ltd. Co.). The above-mentioned abbreviations widely exist in documents and cannot be matched to their corresponding full names directly. Secondly, the abbreviations bring about severe ambiguity problems. On one hand, an abbreviation can be composed of arbitrary characters (even a common Chinese word), which may appear frequently in the documents, but does not actually mean to a CON. For instance, for the name "上海扩大贸易有限公司" (Shanghai Kuoda Trade Ltd. Co.), "扩大" (Kuo-da) can be an abbreviation, but "扩大" is a common verb in Chinese. We should avoid linking this kind of words as far as possible. On the other hand, several CONs may share the same abbreviation. E.g. "上海安吉广告有限公司", "上海安吉家具有限公司" (Shanghai Anji Furniture Ltd. Co.) and "安吉汽车租赁有限公司" (Anji Car Rental Ltd. Co.) can all be abbreviated as "安吉公司" or "安吉". For an abbreviated mention like this in the documents, we need to decide which full name in the given list it refers to. All these obstacles make our task difficult in case of the list-like KB.

Figure 1 shows the brief workflow of our proposed approach. It contains an offline step and an online step. The offline step enriches the KB by the possible abbreviations beforehand. For the input of KB, we describe the steps with the example "上海安吉广告有限公司". The segmenter & annotator component segments the full name of an organization name into parts and annotates each part with predefined types (suffix, region, core and modifier). For the example, it is segmented into "上海" (Shanghai), "安吉" (Anji), "广告" (Ad), "有限" (Ltd.), "公司" (Co.), among which "上海" is a region word, "安吉" is the core, "广告" and "有限" are modifiers, and "公司" is the suffix. Observing that the vocabularies of regions, modifiers, and suffixes for organizations are relatively finite, we leverage external KBs (Hudong Baike[1] and Sogo Cell Word Stock[2]) to collect them. After the full name is processed into the annotations, the abbreviation generator component generates its possible abbreviations, such as those of "上海安吉广告有限公司" mentioned previously. The online step aims to instantaneously generate and disambiguate links between documents and the KB. After a mention is detected from the documents, a set of candidate organization names from

---

[1] http://www.baike.com/

[2] http://pinyin.sogou.com/dict/

the KB are generated by the "candidate finder" component. Like the mention "安吉", several CON entities from the KB appear in the candidate set. To determine the real linked name as well as eliminate the ambiguities, we propose a two-stage approach, which is implemented in the component of disambiguator. The first stage determines the link of relatively full name mentions, and the second stage determines the link of abbreviated names under the constraints of co-occurrence of the full name and abbreviation in the same document or the document cluster that may describe the same organization. The detailed description of our approach is presented in the following sections.

We perform our approach on police inquiry records in Shanghai, and the KB is a list of organizations, most of which are situated in Shanghai. We get 95.17 % accuracy in CON annotation, which outperforms the results of the methods based on the segmentations of the existing tools. The experiments of linking indicate our proposed two-stage linking approach performs significantly better than a one-stage method.

The remainder of the paper is organized as follows. In Sect. 2 we introduce our method of generating abbreviations of organizations. Section 3 describes the proposed linking generation approach. Section 4 includes the details of experiments. Section 5 lists related work and we conclude the paper with some future work in Sect. 6.

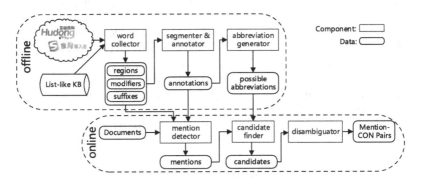

Fig. 1. Workflow of the proposed approach

2 Abbreviation Generation

To enrich the KB, we generate abbreviations by manually defined rules, which are presented in Sect. 2.3. These rules are based on the annotated full names to a predefined pattern, which is presented in Sect. 2.1. In Sect. 2.2, we propose the segmenting and annotating method. The frequent words in CONs are collected through several external KBs at first.

2.1 Pattern Definition of CON

Due to the complexity of a Chinese organization name, researchers have different understanding of the inner structure. Inspired by the works of Zhong et al. [1] and

Chua et al. [2], an organization is usually composed of four parts: namely region, core, modifier and suffix. The CON pattern is defined as follows:

<region> + ... + <region> + <core> + <modifiers> + ... + <modifier> + <suffix>

Table 1 gives several examples. We use "r" to denote a region word, "o" to denote a core word, "m" to denote a modifier word, and "s" to denote a suffix.

"Core" is usually the core part among the four components, and can often become an abbreviation. It is restricted that if there exists a core word, it should be segmented to a whole word. A modifier often represents a type of industry. A CON may contain multi region prefixes (the second example in Table 1) or multi modifiers (the third one). Suffix is a single word, too. Some organization names may have no region word (the forth example in Table 1), no modifier word (the first example), or no core name (the second example). The last example gives a nested name. Note that the order of these four parts are relatively fixed. The region part (if there exists) is always at the beginning of the name, followed by the core part, the modifiers, and the suffix.

Table 1. Examples of the defined structure of CON

CON	Structure
上海金海湾浴室 (Shanghai Jinhaiwan Bathroom)	上海(Shanghai)/r_1 金海湾 (Jinhai-wan)/o 浴室(Bathroom)/s
上海普陀区中医医院 (Shanghai Putuo Chinese Medicine Hospital)	上海(Shanghai)/r_1 普陀区(Putuo)/r_2 中医 (Chinese Medicine)/m_1 医院(Hospital)/s
讯和科技发展有限公司 (Xunhe Technology Development Ltd.)	讯和(Xunhe)/o 科技(Technology)/m_1 发展 (Development)/m_2 有限(Ltd.)/m_3 公司(Co.)/s
上海冠盛饰品有限公司马桥分厂 (Guansheng Jewelry Ltd. Co. Maqiao Branch)	上海(Shanghai)/r_1 冠盛(Guansheng)/o 饰品 (Jewelry)/m_1 有限(Ltd.)/m_2 公司(Co.)/m_3 马 桥(Ma-qiao)/m_4 分厂(Branch)/s

2.2   CON Segmenting and Annotating

Collecting organization specific words. We firstly collect common region, suffix and modifier words to help segmenting and annotating.

Region words. They can be easily obtained from Sogou Cell Word Stock. We download and employ three dictionaries: (1) Countries and regions among the world; (2) Provinces, cities, districts and counties in China; (3) Town, village and street names in main cities in China.

Suffix words. We observe that the existing Chinese NLP tools perform well in segmenting the suffix part of the CON. Hence, in order to collect the common suffix words, we segment all the names in the organization name list using NLPIR 2014[3]

---

[3] http://ictclas.nlpir.org/

(Sect. 4.1 shows that it performs best compared with other tools), and count the last words. Considering that NLPIR may produce some incorrect segmentations and bring about noises, we eliminate the low frequent words.

Modifier words. Collecting modifier words is complicated and consists of 3 steps.

Firstly, we use Hudong Baike, a Chinese Knowledge Base, which contains information of a large amount of organizations. The infobox corresponding to an organization usually contains properties such as "Type of Industry" and "Products", the values of which often involve several words that can be the modifiers in a CON. Therefore, we crawl the infoboxes of these organizations. From the values of the properties, we extract and collect high-frequency words that separated by enumeration comma " 、 ", and limit the length of words to 2 or 3 characters.

Secondly, we count the high-frequency words in the existing organization list. In this step, we simply use existing tool to segment the names, and extract words in the middle. The collected region words are added to the user-dictionary of NLPIR. It is restricted that the counted words must be situated at the position that has an interval of more than one character to the end of the region prefix.

NLPIR, however, cannot split the words perfectly. Some modifier words are separated to individual characters because of the insufficiency of its dictionary. E.g. "足部" (foot) and "木业" (wood) are split into individual characters "足", "部" and "木", "业" respectively. To obtain the out-of-dictionary words, we count common character subsequences among the segmented full names in the KB. Continuous single characters that appear in high frequency are grouped together as modifiers.

Segmenting. The existing Chinese NLP tools, such as NLPIR, Ansj,[4] and FNLP,[5] have a satisfactory performance when handling news, articles and other kind of ordinary texts. But experiments show that they perform relatively poor in segmenting the CONs. Although the problem of dictionary limitation can be resolved by importing collected new words into the user-dictionary, certain problems remain unsolved in our CON segmentation task: (a) There are a large amount of common words in ordinary text in the dictionary, so that some ambiguities caused by these cannot be resolved properly; (b) the granularity is not consistent. Continuous short words may be segmented to a whole longer word sometimes; (c) the Named Entity Recognition (NER) implemented by them may lead to some coarse-grained results and wrong segmentations. A more detailed and experiment based analysis is given in Sect. 4.1.

Given these concerns, we implement a segmentation method, specific to CONs. The method is based on Bi-Direction Maximum Match (BDMM). The dictionary we use in BDMM only contains the words that we collect, so that the coarse-grained words and the ambiguities that involve common words are unlikely to appear. Besides, since the BDMM is just a mechanical segmentation method without any additional functions, the problem of NER in the existing tools will not exist.

---

[4] https://github.com/ansjsun/ansj_seg
[5] https://github.com/xpqiu/fnlp/

Annotating. On the basis of segmentation, we can simply tell apart the suffix, which is in fact the last word. Then the annotation task separates the other three parts by determining the boundaries, among regions, core and modifiers.

We have already obtained nearly sufficient amount of region words and modifier words. A natural idea is to match the segmented full name from the beginning with the region collection to determine the end of the region part. The core and modifier parts can be separated through matching the words after the region in the sequence one by one to our modifier word collection.

However, some issues surfaced in the process. On one hand, ambiguities may appear between region and core parts. The core word in a CON may be same to certain region name. Thousands of street names are added into the modifier list, causing this kind of ambiguities not to be ignored. For instance, "上海富民工艺品雕刻厂" (Shanghai Fumin Object Carve Factory), "富民" (Fumin) is the core part, while there is a town named "富民". On the other hand, a modifier word may in fact be a part of the core of a CON. For example, for the name "兆能源酒店设备用品有限公司" (Zhaonengyuan Hotel Facility Ltd. Co.), the core word is "兆能源" (Zhaonengyuan), while "能源" (Nengyuan, means energy) can be a modifier in some other names.

Following the discussion above, we add some additional criteria to distinguish the boundary between the region part and the core part, as well as the boundary between the core part and the modifier part, based on the observation that a core word usually contains at least 2 characters. After we match the continuous words that belong to region collection from the beginning of the full name, and the first word that exists in modifier collection after region words, there are three possible situations:

- An interval of two or more than two characters exists between the last continuous region words and the first modifier words. The characters are considered as the core part, and the boundaries are determined. Such as the name "上海安吉广告有限公司", the region part is "上海", and the first matched modifier word is "广告". Then "安吉" is the core part.
- Only one character exists between region and modifier parts. In such case, it is necessary to decide whether the last matched region word or the first matched modifier word should be merged into the core word. If a segmentation satisfies either of the two conditions, we put the last matched region word into the core: (a) The last matched region word does not refer to a city, a country or a district; the last matched region word does not end with a "key character" of region name, such as "镇" (town), "路" (road). (b) The last region word does not belong to the former region in the segmentation. The first matched modifier word is annotated as a modifier. If none of these conditions is satisfied, the first matched modifier is merged into the core. Using these strategies, for the name "上海新明星物业管理有限公司" (Shanghai Xinmingxing Property Management Ltd. Co.), although "新明" (Xinming) is a street name, "新明星" (Xinmingxing) is annotated as the core part.
- No other characters exist between matched region and modifier parts. The strategy is similar to that in the previous situation, except that when neither of the two conditions is satisfied, we consider the CON has no core word. Then there is no core

word in "上海虹口人力资源有限公司" (Shanghai Hongkou Human Resource Ltd. Co.): "虹口" (Hongkou) is a region, "人力" (Human Resource) is a modifier.

2.3 Generating Abbreviations

Rules to generate abbreviations are defined based on annotations. More specifically, a generated abbreviation is composed of some of the four parts from the defined CON pattern. The generation rules are as follows: (1) For an organization name that contains core word, we assume that it is already annotated in the form of $r_1, r_2, \ldots, r_{N_r}, o, m_1, m_2, \ldots, m_{N_m}, s$, where $r_i (0 \leq i \leq N_r)$ is a region word, $m_i (0 \leq i \leq N_m)$ is a modifier word, $N_r, N_m \geq 0$, o and s are core word and suffix respectively. The rules of abbreviation are listed in Table 2 with some examples. (2) For a name without core word, annotated as $r_1, r_2, \ldots, r_{N_r}, m_1, m_2, \ldots, m_{N_m}, s$, the rules are listed in Table 3. Some generated abbreviations are not reasonable, for example, "农业银行有限", but most of this kind of phrases are unlikely to appear in documents. In the tables, the parentheses mean that the corresponding parts may or may not exist.

Table 2. Rules of abbreviation for a CON with core name

Pattern	Example
Full name annotation	上海乌龙网络技术发展有限公司 / Shanghai Wulong Network Technology Development Ltd.
$o, (s)$	乌龙(公司) / Wulong (Co.)
$o, m_{i_1}, m_{i_2}, \ldots, m_{i_k}, (s)$ $(i_1 < \ldots < i_k)$	乌龙技术发展(公司) / Wulong Technology Development (Co.)
$r_{i_1}, \ldots, r_{i_k}, o, (s)$	上海乌龙(公司) / Shanghai Wulong (Co.)
$r_{i_1}, \ldots, r_{i_k}, o, m_{i_1}, \ldots, m_{i_k}, (s)$	上海乌龙网络(公司) / Shanghai Wulong Network (Co.)

Table 3. Rules of abbreviation for a CON without core name

Pattern	Example
Full name	中国农业银行股份有限公司 (Agricultural Bank Of China Ltd. Co.)
$m_1, m_2, (s)$	农业银行(公司) / Agricultural Bank (Co.)
$m_1, m_2, m_{i_1}, \ldots, m_{i_k}, (s)$	农业银行有限(公司) / Agricultural Bank Ltd. (Co.)
$r_{i_1}, \ldots, r_{i_k}, m_1, m_2, (s)$	中国农业银行(公司) / Agricultural Bank Of China (Co.)
$r_{i_1}, \ldots, r_{i_k}, m_1, m_2, m_{i_1}, \ldots, m_{i_k}, (s)$	中国农业银行有限(公司) / Agricultural Bank Of China Ltd. (Co.)

3 Linking Generation

3.1 Mention Finding and Candidate Finding

For efficiency consideration, we do not directly use the enriched KB (i.e. all the generated abbreviations) to find mentions. Instead, we detect mentions by finding continuously appeared key words – the common region words, modifiers, suffixes, as well as all the core words annotated in the KB. We restrict that a mention should be composed of one core word or more than one word of any kind. The mentions found in this way may contain many noises, which can be excluded through our proposed linking determination method.

For every detected mention, we find the organization names from the KB that may be linked as candidates. It is notable that we do not consider the left boundary and right boundary of the mentions while finding them. Hence, when judging whether an organization name in the KB is a candidate for a mention, we only need to ensure that an abbreviation of an organization is a substring of the mention.

3.2 Linking Determination

It is attributed to two aspects of reasons that multiple entities in the KB appear in the candidate set. Firstly, some full names are inevitably incorrectly segmented and annotated, and the wrong abbreviations happen to match the mention. Secondly, several names may share a common abbreviation, as discussed in Sect. 1. Even for a mention of full name, sometimes a company may have branch companies, and they can all take the company name as an abbreviation. Then all the branch companies are added to the candidate set.

To filter out the true linked entity from the candidates, our ranking strategy considers two measures successively. Firstly, for the candidates of a mention, we consider their respective longest abbreviations that match the mention as the best solution. The longer the matched abbreviation is, the higher the corresponding candidate ranks. Secondly, for the candidates whose longest matched abbreviations have the same length, we rank them by the length of the full names themselves increasingly. In other words, the shorter full name ranks higher. To implement these strategies, we define the score for a candidate as Eq. (1):

$$score(m, c) = K \cdot longestMatchedAbbreviation - fullNameLength \qquad (1)$$

Where m is a mention, c is a candidate corresponding to m. K is a sufficiently large constant, such that there is no full organization name whose length is larger than K. The candidates are ranked by the score decreasingly.

As discussed above, two aspects of ambiguities are caused by the existence of abbreviations. However, the inquiries themselves should not have these ambiguities while being read. Thus, we believe that for the names that may have ambiguities, especially the abbreviated ones, a full name is usually mentioned in the same document or a document cluster referring to the same organization. We use co-occurrence of the abbreviations and their corresponding full names to resolve these problems.

The determination process contains two stage. Because a modifier can indicate the type of industry, we reckon that it can disambiguate organization names that share the same core word. Consequently, in first stage, after picking out the top-ranked candidate for a mention, we judge whether the longest abbreviation that matched the mention contains a modifier word. If it does, the link is then determined. This kind of names is recognized as relatively full names. For those mentions which are not determined, we process them in the second stage, in which the ranking procedure only consider the candidates that have been linked to in the documents describing the same CON in the first stage, so that the ambiguities can be eliminated. Specifically, we use the results of the first stage to restrict the abbreviated names' links. For example, we can determine that "安吉" represents "上海安吉广告有限公司" because the full name has appeared in the same document. "扩大" will not be linked because there is no full name, which can be abbreviated as "扩大" has been linked in the same document cluster. The ranking method is also used in the second stage. Figure 2 lists the pseudo-code of determining links in the documents that may refer to the same CON.

Algorithm 1. Determine links of documents that refer to a same CON

Input: $\{Mention_i\}$ Set of mentions in documents that may refer to a same CON; $\{Candidate_i\}$
Set of candidates for $Mention_i$, where, for each i, $Candidate_i = \{Candidate_{i,j}\}, 1 \leq j \leq N_i$

Output: (mention, entity) pair list

1: $PairList \leftarrow Empty$
2: $AbbrevMentionSet \leftarrow Empty$
3: $LinkedCandiSet \leftarrow Empty$
4: **for** each $m \in \{Mention_i\}$ **do**
5: $\quad c \leftarrow \underset{Candidate_{i,j}}{\arg\max} \ \{score(m, Candidate_{i,j}) \mid 1 \leq j \leq N_i\}$
6: $\quad$ **if** (m, c) is a link of relatively full name **then**
7: $\quad\quad$ put (m, c) into $PairList$
8: $\quad\quad$ put c into $LinkedCandiSet$
9: $\quad$ **else**
10: $\quad\quad$ put $Mention_i$ into $abbrevMentionSet$
11: $\quad$ **end if**
12: **end for**
13: **for** each $m \in AbbrevMentionSet$ **do**
14: $\quad c \leftarrow \underset{Candidate_{i,j}}{\arg\max} \ \{score(m, Candidate_{i,j}) \mid 1 \leq j \leq N_i, Candidate_{i,j} \in LinkedCandiSet\}$
15: $\quad$ **if** c exists **then**
16: $\quad\quad$ put (m, c) into $PairList$
17: $\quad$ **end if**
18: **end for**
19: **return** $PairList$

Fig. 2. Pseudo-code of link determination

4 Experiments

4.1 Performance of Segmenting and Annotating

In our experiments, the KB is a list of Shanghai organization names provided by Shanghai Public Security Bureau, which contains totally 1,941,956 Chinese organization full names, after eliminating duplicates. It covers a variety of organizations, including (branch) companies, stores, banks and so on.

There are no ground truths available for segmentations and annotations of these organizations. So we randomly select 600 names from the KB as samples to test the performance of the proposed method. The results are manually labeled and counted.

Several labeling criteria are specified: (a) If a region word is splitted, label it as wrong; (b) If a modifier word is splitted, label it as wrong; (c) If there exists a word (unless it is a complete core word) which contains more than one complete word, label it as wrong; (d) If the core part is splitted, omit it.

Performance of NLP Tools. In this section, we have a detailed test to the performance of some existing NER tools, including Ansj, NLPIR and FNLP. For every tool, we use it to segment the samples with its default dictionary and with the supplementary of region words separately. These experiments aim at choosing the best way to help to collect suffix words and modifier words, as we introduced in Sect. 2.2. Besides, they act as baseline methods compared to our proposed one. Table 4 shows the results.

Table 4. Segmentation performance of existing tools

Tool	Dictionary	Number of wrongs	Accuracy
Ansj	Default	84	86.00 %
	Add region words	63	89.50 %
FNLP	Default	57	90.50 %
	Add region words	57	90.50 %
NLPIR	Default	62	89.67 %
	Add region words	**47**	**92.17 %**

Ansj performs worst compared to the other two (86.00 %). The most serious problem is its poor ability to solve ambiguities. More precisely, while confronting a character sequence $c_1c_2c_3$, if c_1c_2 and c_2c_3 are both in dictionary, Ansj segments the sequence as c_1, c_2c_3. "上海洋光经贸公司" (Shanghai Yangguang Economy and Trade Ltd.), for instance, Ansj segments it to "上, 海洋, 光, ..." ("Shang, haiyang, guang, ...", where "haiyang" means sea). After adding region words to its dictionary, these mistakes still exist.

For FNLP, the results of the two dictionaries are the same (90.50 %), better than the other two tools with default dictionaries, because its default dictionary covers region words. But the problem caused by Named Entity Recognition appears much. For example, "南铁工贸" (Nantie Industry and Trade) and "捷豪钢" (Haojie Steel) in "上海南铁工贸有限公司" (Shanghai Nantie Industry and Trade Ltd. Co.) and "上海捷豪钢结构工程有限公司" (Shanghai Haojie Steel-structure Engineering Ltd. Co.), are respectively recognized as an organization and a person name. Besides, many coarse-grained words exist, such like "经纪有限" (Agency Limited).

NLPIR with dictionary with region words performs best (92.17 %). Thus, while collecting modifier words and suffix words, we use NLPIR to segment the CONs.

Proposed Method. The final results of segmenting and annotating are listed in Table 5. Three experiments on NLPIR with different dictionaries are performed, aiming to prove the effect of our collected words. The line of "NLPIR + region" means the dictionary is complemented with region words. The second line "NLPIR + region + Baike" means the dictionary contains region words and the words extracted from Hudong Baike (totally 1561 ones). "New words" in the third line means the words extracted through common character sequences presented in Sect. 2.2 (301 ones).

Table 5. Final results of segmentation and annotation

Method	Segmentation		Annotation	
	Wrong	Accuracy	Wrong	Accuracy
NLPIR + region	47	92.17 %	56	90.67 %
NLPIR + region + Baike	39	93.50 %	50	91.67 %
NLPIR + region + Baike + new words	20	96.67 %	42	93.00 %
Proposed method	**14**	**97.67 %**	**29**	**95.17 %**

Every time an additional collection of words is added, the accuracies of segmenting and annotating using NLPIR increase by more than 1 percentage, which proves the each step of collecting words to be valuable. The insufficient modifier word collection, ambiguities between region, core and modifier parts may all produce errors.

Our BDMM-based method resolves most problems caused by segmentation granularity, ambiguities and NER. Compared to the best result using NLPIR, accuracy of segmentation increases 1 percentage, and accuracy of annotation increases 2.17 percentage. Therefore, our method is more effective when processing CONs. The remaining problems are mainly owning to the deficiency of dictionaries, so that some low frequent modifier words could not be segmented or annotated correctly.

4.2 Performance of Our Linking Generation Approach

We use police inquiry documents as corpus to test our linking generation approach, which are of real cases in Shanghai occurred during the year of 2012–2013. Police inquiries are written in the form of questioning and answering. They usually involve many organization names related to the cases. For convenience, the organization names are often written as abbreviated form. In one case, there are often more than one person is questioned, and a person is sometimes questioned more than one time. So there are often more than one inquiry documents corresponding to one case. For a specific case, organization names related to it often appear multiple times in the corresponding inquiries. So we consider the co-occurrence of full names and abbreviations in inquiry documents of a case and apply our linking approach to them.

Totally 206 inquiry documents participate in the experiment. They are corresponding to 37 different cases respectively. There are totally 7,651 questions and answers, containing 21,657 sentences. We set $K = 100$ in Eq. (1).

To prove the effectiveness of the two-stage approach, we use a one-stage method as a baseline. In this method, for every mention, we directly rank the candidates of it and choose the first one as the linked entity, without considering whether the mention is a full name or an abbreviation.

Table 6 gives the evaluation results, including (micro) precision, recall and F-measure. In the table, the term "Recall of linking" means that the percentage of correctly linked mentions to all the mentions in the documents which have target entities to link in KB. And the term "Recall of recognition" means the percentage of correctly linked mentions to all the organization names appeared in the documents, regardless of whether there are entities in KB to link to. Besides, we consider the repeatedly appeared mentions. The term "Count repeats" distinguishes whether the results consider the repeatedly appeared mentions. "Yes" means that it counts all mentions regardless of their repeats; "No" means that the duplicated mentions are omitted. We list both the number of correct links and the corresponding percentage.

Table 6. Results of the linking approach, two-stage vs. onestage (Micro)

	Count repeats	One-stage		Two-stage	
		Number	Percentage	Number	Percentage
Precision	Yes	1343/30452	4.41%	2401/2670	89.93%
	No	243/3459	7.03%	284/332	85.54%
Recall of linking	Yes	1343/2606	51.53%	2401/2606	92.13%
	No	243/324	75.00%	284/324	87.65%
Recall of recognition	Yes	1343/3217	41.75%	2401/3217	74.63%
	No	243/664	36.60%	284/664	42.77%
F-Measure of linking	Yes		0.0813		0.9102
	No		0.1285		0.8658

Table 7. Results of linking in each stage, in the two-stage method

	Count repeats	First stage		Second stage	
		Number	Percentage	Number	Percentage
Precision	Yes	791/890	88.88 %	1610/1780	90.45 %
	No	183/219	83.56 %	97/109	88.99 %

Through the comparison between the one-stage method and the two-stage method, we can see that the latter one significantly outperforms the former one. For one-stage method, the precision is pretty low (4.41 % and 7.03 % for "count repeats" and not "count repeats"). While the precisions of the two-stage method are 89.93 % (count repeats) and 85.54 % (not count repeats). These results indicate that the ambiguity problems induced by abbreviations are very common and make the performance of one-stage method poor, but our proposed method can resolve them effectively.

A detailed result of each stage in the two-stage method is listed in Table 7. There are totally 1610 mentions (97 without repeats) linked to entities in second stage, which account for a considerable proportion.

Recall of the two-stage method is also larger than that of one-stage method, owning to the much more correctly linked abbreviations. Then the F-Measure of our two-stage method is much higher. Recall of recognition is not that high, because many CONs in the inquiries do not exist in the KB (i.e. no entity to link).

Some CONs in the inquiry documents are written in abbreviated form directly. Since there are not any relatively full names appear, the corresponding CON entities do not exist in the candidate set in the second stage, so that these mentions cannot link to their corresponding entities in KB successfully. However, in the one-stage method, some of them can link to the correct entities, if the entities rank first by chance. This phenomenon explains that the gap of recalls between the one-stage method and the two-stage method is not that large as the gap of precisions.

5 Related Work

There are many studies focusing on recognizing Chinese organization names, especially their recognition. There are two kind of methods, rule-based methods and statistic-based methods. Wang et al. [3] presented a simple rule-based approach to recognize CONs. They detected left and right boundaries and used a length constraint and POS-tag constraints to determine further. Ling et al. [4] took advantage of various types of features to address CONs, including boundaries, behaviors and structure patterns. Similar to other rule-based approaches, they always consider the suffixes as right boundaries, so they cannot recognize abbreviations without suffixes. Besides, they all utilize POS-tag generated by segmentation tools to construct rules, which is unreliable, especially for the inner structure of CONs. Many studies use Conditional Random Fields (CRFs) to accomplish the Chinese NER task. In most recent studies, Fu et al. [5] proposed a dual-layer CRFs based method to recognize nested named entity; And Wu et al. [6] combined CRFs with dynamic gazetteers to become more adaptive to new domains and new named entities. CRF-based methods often need large amount of labelled data. The varying lengths of CONs and the diverse characters in core words affect the effectiveness of these methods.

Recently, EL has got various studies. Most of them use Wikipedia as the KB and focus on the disambiguation process. Some representative studies include the work of Zhang et al. [7], Han et al. [8], Liu et al. [9], Shen et al. [10]. Zhang et al. made use of additional information sources from Wikipedia, and presented a novel method to generate corpus annotation, then train a binary classifier reduce the ambiguities. Han et al. proposed a graph-based method, based on which they can jointly infer the referent entities of all mentions by exploiting the interdependence. Liu et al. studied the task of Entity Linking for tweets. To address the challenges of the dearth of information in a tweet and the rich mention variations, they proposed a collective inference method that simultaneously resolves a set of mentions. Shen et al. referred the entities in Web lists to the knowledge base, with the assumption that the entities mentioned in a Web list tend to be a collection of entities of the same conceptual type. The entities in the Web lists have no context, while in our work, the entities in KB have no context.

6 Conclusion and Future Work

In this paper, we link mentions in texts to a list-like KB of CONs. To resolve the problems brought by various abbreviations, we designed specified segmenting and annotating methods, using the words collected via various ways. We applied a BDMM-based method to accomplish the segmenting task, and performs better than other public NLP tools. In the course of linking, the co-occurrence of full names and abbreviated ones in the same case were utilized to disambiguate the linking of abbreviations, getting a satisfying precision and recall.

Recently, Chinese organization recognition is still a difficult problem. Although it has been widely studied, the existing methods and tools do not perform very well on arbitrary documents, especially for the abbreviations. Based on our work, we can explore distant supervision method for the NER task. That is to say, we can make use of the results of linking to help the recognition. An idea is treating the high confidence linked mentions as labeled sequences, to train a statistical learning model or construct a heuristic rule based method.

Acknowledgements. This work is funded by The 3rd Research Institute of The Ministry of Public Security through project No: C13601. We thank Tong Ruan for the guidance of the project, and thank Chen Wang for her proofreading.

References

1. Zhong, L.W., Zheng, F.: Study on approach to retrieval of chinese organization name based on its abbreviated name. J. Chin. Inf. Process. **21**, 38–42 (2007)
2. Chua, T.S., Liu, J.: Learning pattern rules for chinese named entity extraction. In: Proceedings of AAAI/IAAI, 411–418 (2002)
3. Houfeng, W., Wuguang, S.: A simple rule-based approach to organization name recognition in chinese text. In: Gelbukh, A. (ed.) CICLing 2005. LNCS, vol. 3406, pp. 769–772. Springer, Heidelberg (2005)
4. Ling, Y., Yang, J., He, L.: Chinese organization name recognition based on multiple features. In: Chau, M., Wang, G., Yue, W.T., Chen, H. (eds.) PAISI 2012. LNCS, vol. 7299, pp. 136–144. Springer, Heidelberg (2012)
5. Fu, C., Fu, G.: A dual-layer CRFs based method for chinese nested named entity recognition. In: 9th International Conference on Fuzzy Systems and Knowledge Discovery, pp. 2546–2550. IEEE, New York (2012)
6. Wu, X., Wu, Z., Jia, J., et al.: Adaptive named entity recognition based on conditional random fields with automatic updated dynamic gazetteers. In: 8th International Symposium on Chinese Spoken Language Processing, pp. 363–367. IEEE, New York (2012)
7. Zhang, W., Su, J., Tan, C.L. et al.: Entity linking leveraging: automatically generated annotation. In: COLING 2010, pp. 1290–1298. ACL, Stroudsburg (2010)
8. Han, X., Sun, L., Zhao, J.: Collective entity linking in web text: a graph-based method. In: 34th ACM SIGIR International Conference on Research and Development in Information Retrieval, pp. 765–774. ACM, New York (2011)
9. Liu, X., Li, Y., Wu, H., et al.: Entity linking for tweets. In: The 51th Annual Meeting of the Association for Computational Linguistics, pp. 1304–1311. ACL, Stroudsburg (2013)
10. Shen, W., Wang, J., Luo, P., et al.: LIEGE: link entities in web lists with knowledge base. In: The 18th ACM SIGKDD International Conference on Knowledge Discovery and Data Mining, pp. 1424–1432. ACM, New York (2012)

Chinese Textual Contradiction Recognition Using Linguistic Phenomena

Maofu Liu[1,2], Yue Wang[1,2(✉)], and Donghong Ji[3]

[1] College of Computer Science and Technology, Wuhan University of Science and Technology, Wuhan 430065, China
liumaofu@wust.edu.cn, 289455785@qq.com
[2] Hubei Province Key Laboratory of Intelligent Information Processing and Real-time Industrial System, Wuhan 430065, China
[3] School of Computer, Wuhan University, Wuhan 430072, China
dhji@whu.edu.cn

Abstract. Detecting contradictive texts is a crucial and fundamental work for text understanding just like textual entailment. Textual contradiction occurs when two different texts cannot both be true at the same time. This paper focuses on the linguistic phenomena behind textual contradiction, including quantity exclusion, temporal exclusion, spatial exclusion, modifier exclusion, antonym and negation. In this paper, the Chinese textual contradiction approach using linguistic phenomena has been put forward and a number of experiments on basis of one textual entailment system have been made to evaluate this approach. The experiment results demonstrate the effectiveness and feasibility of the Chinese textual contradiction recognition approach using linguistic phenomena.

Keywords: Chinese textual contradiction · Linguistic phenomena · Semantic rules

1 Introduction

Textual entailment (TE) in natural language processing is a directional relation between text fragments. The relation holds whenever the truth of one text fragment follows from another text. That "T entails H" means if the truth of hypothesis H can be typically inferred from text T and is expressed as "T⇒H" [1].

It is an important subtask of recognizing textual entailment to find textual contradiction. In propositional logic, textual contradiction is defined as that there is no possible world in which two texts are both true at the same time, nevertheless, for textual inference, textual contradiction relies on human's understanding of "incompatibility" [2]. Considering the following Example 1, one Chinese contradictory text pair, which may not match the strict logical definition of contradiction, fits human intuitions perfectly.

Example 1:
 T1：张三卖给李四一本书。
 H1：李四卖给张三一本书。

© Springer-Verlag Berlin Heidelberg 2014
D. Zhao et al. (Eds.): CSWS 2014, CCIS 480, pp. 111–122, 2014.
DOI: 10.1007/978-3-662-45495-4_10

Early RTE (Recognizing Textual Entailment) tasks required participants to determine whether or not a given text entails another text, which is a typical binary classification task. Since RTE-3, another two relation types, contradiction and independence, have been included [3]. The RITE (Recognizing Inference in TExt) task in the 10th NTCIR (NII Test Collection for IR Systems), further extended the entailment relation types to four, including forward entailment, meaning T entails H while H does not entail T, bidirectional entailment, which denotes T and H can infer each other, contradiction, which means T and H generally cannot be true at the same time, and otherwise independence, in the given text pair (T, H) [4].

As the participant of RITE-2 task at NTCIR-10, initially we implemented a textual entailment recognition system based on support vector machine, using statistical features, lexical semantic features and syntactic features [5]. The official evaluation results indicated that our system had a poor performance on contradiction recognition. This observation motivated us to do analysis on contradictory text pairs to investigate the linguistic phenomena behind textual contradiction. Furthermore, a Chinese textual contradiction recognition approach based on linguistic phenomena has been proposed in this paper.

Since this paper holds the hypothesis that there would be large room for improving contradiction detection in our system, we simplify the multiple classification task of NTCIR-10 RITE-2 as a binary classification problem in order to focus on textual contradiction. So the other three relation types including forward entailment, bidirectional entailment and independence have been marked as the label "Non-Contradiction". The experimental results demonstrate the effectiveness of the textual contradiction recognition approach, making the improvement of the textual contradiction recall from 0.283 to 0.547.

2   Related Work

So far, most of the work in text understanding has focused on textual entailment and only a little of attention has been paid to textual contradiction detection. Crouch et al. [6] first emphasized the importance of contradiction detection for text understanding, which should be considered as vital as entailment. Later, Harabagiu et al. [7] first reported experimental results on contradiction identification and proposed a framework for recognizing contradictions by extracting negation, antonym and contrast relations. Ritter et al. [8] gave a deep analysis of the RTE-3 extended task and presented a case study of contradiction detection based on functional relations. Marneffe et al. [9] presented a typology of contradictions and provided two primary types of contradictions, the contradiction arising from negation, antonym, or numeric mismatch and the contradictions resulting from factive or modal words, text structural and subtle lexical contrasts, and background knowledge. Some useful features were also proposed to identify contradiction in texts such as antonym, negation, fact, relations, modality, and structures. Voorhees [10] gave an analysis of RTE-3 extended task and focused on contradictions that lacked background knowledge. Shih et al. [11] aimed to address the problem of the shortage of specific background knowledge in contradiction detection by using a search engine result returned by well designed query strings.

The previous work above mainly focused on contradictions in English text pairs. In this paper, we aim at Chinese textual contradictions and deeply mined the contradictory linguistic phenomena. Moreover, we adopt statistical features to calculate text similarity for the reason that high textual similarity should be considered as the essential premise of textual contradiction. The semantic rules according to the linguistic phenomena are also introduced to improve the accuracy of contradiction recognition.

3 Textual Contradiction Recognition Approach Using Linguistic Phenomena

3.1 Model Overview

In our approach, contradictory text pairs are detected in five steps and model overview is shown in Fig. 1.

In the data preprocessing, the main work of the system is to segment the Chinese words, remove the stop words and parse the text pairs. For testing dataset, tagging and named entity recognition are also needed besides the above steps, which is prepared for the extraction of contradiction related information.

In feature extraction stage, three kinds of features, statistical features, lexical semantic features and syntactic features, will be extracted. Statistical features can be classified into two categories, word set features and vector features. Vector features, such as Manhattan distance, Euclidean distance, cosine similarity, are calculated according to TF*IDF. Lexical semantic features are calculated to evaluate the semantic similarity of the text pairs by using some semantic resources such as Hownet and TongyiCilin. Syntactic features are extracted by parsing the text pairs.

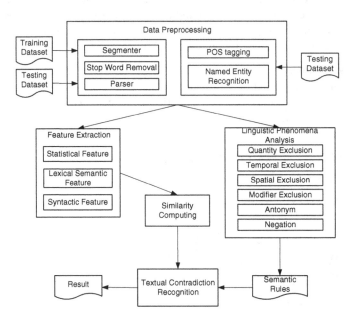

Fig. 1. Model overview

After extracting the text pair features, the overall similarity of Chinese text pairs is calculated because the mismatching information is the strong evidence for the textual contradiction in many situations, especially when the text pair holds high similarity. In this stage, textual similarity must be considered as the essential premise of textual contradiction. The overall similarity of T and H is evaluated by the weighted average of text similarity like cosine similarity, Hownet similarity, word overlap and ratio of the same words extracted in the feature extraction stage, which is defined in the following formula (1), where Si and Wi refer to each kind of similarity and its weight respectively.

$$Similarity = \frac{\sum S_i * W_i}{n} \tag{1}$$

And then, the Chinese contradiction linguistic phenomena are deeply mined by analyzing the Chinese contradictory text pairs. A typology of Chinese contradiction linguistic phenomena is presented, including quantity exclusion, temporal exclusion, spatial exclusion, modifier exclusion, antonym and negation. According to various linguistic phenomena mentioned above, this model generates the corresponding semantic rules.

Finally, the textual contradiction will be recognized using the Chinese semantic rules, including quantity rule, temporal rule, spatial rule, modifier rule, antonym rule and negation rule, generated by the linguistic phenomena analysis.

3.2 Contradictory Linguistic Phenomena and Semantic Rules

In NTCIR-10 RITE-2 task, our initial system with statistical features based on SVM classifier performed poorly on Chinese textual contradiction recognition. In order to detect Chinese textual contradiction successfully, it is necessary to have a deep analysis on the linguistic phenomena behind contradictory text pairs. In this paper, we provide six categories of linguistic phenomena related to textual contradiction in the RITE-2 dataset, which are given in the following subsections.

(1) Quantity Exclusion

Quantity exclusion is defined as a numeric mismatch between T and H. The following Example 2 illustrates various kinds of numeric mismatches which cause textual contradictions between T and H.

Example 2:
T2：艾弗森职业生涯最高单场得分60分。
H2：艾弗森的最高得分纪录为65分。
T3：阿诺尔特大花直径最多达3米。
H3：阿诺尔特大花直径能够到达3千米。
T4：平均一天睡10个多小时的人最长寿。
H4：每晚平均睡近10小时的人，寿命最长。
T5：大卫像高4.342公尺。
H5：大卫像高2.5米。

The text pair (T2, H2) shows a value mismatch and the value of number in T2 is "60" while that in H2 is "65". As in the text pair (T3, H3), the numbers share the same value "3" while hold different units "米" and "千米" respectively. Another kind of numeric mismatch is range mismatch as in (T4, H4) and the words "多" and "近", meaning more than and less than, determine opposite ranges of the same number "10". In text pair (T5, H5), there exist a value mismatch and a unit mismatch. After unit conversion from "4.342公尺" in T5 to "4.342米", it is also not equal to "2.5米" in H5.

Example 3:

T6：熊猫体长约180厘米。

H6：熊猫身长能达到1.8米。

T7：北极熊平均年龄30岁左右。

H7：北极熊平均年纪三十岁左右。

However, not every kind of numeric mismatch would lead to a textual contradiction. In text pair (T6, H6), although "180厘米" differs from "1.8米", they are equal to each other after unit conversion. In text pair (T7, H7), as the same number "thirty" can be expressed as "30" and "三十" in Chinese and Arabic ways, the text pair will not be considered as a contradictory one.

The four types of numeric mismatches in Example 2 can obviously lead to textual contradictions while the two types in Example 3 cannot do because there are different forms of expressions for the same number in Chinese texts, for example, "四万", "40000" and "4万" all refer to the same number.

Before textual contradiction judgment, the numbers should be normalized and presented as a triple (value, unit, range) by using Stanford POS tagger [12]. We normalize a number as the Arabic one and the units of measurement should also be standardized. The ranges of number can be determined by some signal words such as 大于 (More than)", 小于 (Less than)", 超过 (Over)" or 不足 (Within)". The numeric mismatch, including value mismatch, unit mismatch and range mismatch, could conclude the textual contradiction if T and H have high similarity. According to the linguistic phenomena of quantity exclusion, the corresponding rules have been designed as follows.

Quantity Rule 1: For a given text pair (T, H), which holds high similarity, if the two numbers in T and H have the same unit and range but different values, it can be justified as textual contradiction.

Quantity Rule 2: For a given text pair (T, H), which has high similarity, if the two numbers in T and H have the same value and range but different units, it can be justified as textual contradiction.

Quantity Rule 3: For a given text pair (T, H) which has high similarity, if the two numbers in T and H have the same value and unit but different ranges, it can be justified as textual contradiction.

Quantity Rule 4: For a given text pair (T, H) which has high similarity, if the two numbers in T and H have the same range but different values and units and they aren't equal to each other after unit conversion, it can be justified as textual contradiction.

(2) Temporal Exclusion

Temporal exclusion means a time or date mismatch which could conclude the textual contradiction between T and H. The following Example 4 shows temporal exclusion.

Example 4:
T8：撒切尔于1992年被册封为终身贵族。

H8：1991年撒切尔得到终身贵族的头衔。

T9：京都议定书为1997年12月在气候变化纲要第三次缔约国大会中通过。

H9：一九九七年十二月第三次缔约国会议中通过"京都议定书"。

T10：2005年7月7日的清晨8点50分伦敦多处地铁站爆炸。

H10：2005年7月7日的伦敦地铁爆炸发生于早上8点50分。

In the example above, the temporal exclusion could occur via the year, month, day or format mismatch. The text pair (T8, H8) is contradictory because the year information, "1992年" and "1991年", is different which is a typical temporal exclusion. However temporal expression mismatch may not conclude a contradiction sometimes because the date or time could be represented in various formats in Chinese. As in text pair (T9, H9), "1997年12月" and "一九九七年十二月" refer to the same temporal information. In text pair (T10, H10), "清晨" and "早上" are different descriptions of "morning".

As a result of diverse expressions of date and time, they should be normalized before contradiction identification. For example, "1990/02/21", "19900221" and "1990年2月21日" will not be considered as temporal mismatch. The Stanford POS tagger is used to extract time or date information in the text pairs according to the labels "/T" and "/NT". A temporal mismatch could lead to a contradiction of a text pair if the structural similarity of two sentences is high. The following rule is designed based on temporal exclusion.

Temporal Rule: For a given text pair (T, H) which has high similarity, if the date or time in T and H has a mismatch after normalization, it can be justified as textual contradiction.

(3) Spatial Exclusion

Spatial exclusion is also crucial for the textual contradiction in the case of the spatial information referring to the same event. In text pair (T11, H11), the textual contradiction results from different locations, "中国" and "日本", which are both involved in the same event "原产". Another situation is that the same location information in different events may also conclude textual contradiction. In text pair (T12, H12), the same location information "江西德安" is involved in two different events, "祖籍" and "出生". The spatial information is extracted by Stanford NER (Named Entity Recognizer) according to the label "/GPE" [13].

Example 5:
T 11：土豆原产于中国。

H11：土豆原产于日本。

T12：袁隆平祖籍是江西德安。

H12：袁隆平出生于江西德安。

According to this linguistic phenomenon, the spatial rules, listed as follows, are used to recognize the spatial contradictory text pairs.

Spatial Rule 1: For a given text pair (T, H) which has high similarity, if the different location information in T and H denoted the same event occurring, it can be justified as textual contradiction.

Spatial Rule 2: For a given text pair (T, H) which has high similarity, if the same location information in T and H involved in different events, it can be justified as textual contradiction.

(4) Modifier Exclusion

The different modifiers for the same thing may create textual contradictions sometimes. The different modifiers "唯一" and "次要" make T13 and H13 conflict with each other. However, if the different modifiers are synonym, hypernym or hyponym ones, the modifier exclusion is not sufficient to indicate a textual contradiction. Taking text pair (T14, H14) for example, it is the bidirectional entailed text pair instead of contradictory one because "丰富" and "大量" are synonyms. Similarly, the text pair (T15, H15) is the forward entailed one as the "葱科植物" is the hyponym of the "草本植物". The following semantic rules illustrate the linguistic phenomena mentioned above.

Example 6:

T13：海底地震造成地层大幅度陷落抬升是引发大海啸的唯一原因。

H13：海底地震造成地层大幅度陷落抬升是引发大海啸的次要原因。

T14：草莓含有丰富维生素C。

H14：草莓含有大量维生素C。

T15：韭菜，属多年生葱科植物。

H15：韭菜，属多年生草本植物。

Modifier Rule 1: For a given text pair (T, H) which has high similarity, if there exists a modifier mismatch which is not a synonym pair, it can be justified as textual contradiction.

Modifier Rule 2: For a given text pair (T, H) which has high similarity, if there exists a modifier mismatch which is not a hypernym or hyponym pair, it can be justified as textual contradiction.

(5) Antonym

The antonym is a very useful cue for textual contradiction as the antonym pairs usually convey oppositional information. The antonym pair "富裕" and "清贫" can lead to the textual contradiction between texts T16 and H16. To calculate the pair number of the antonym in text pair (T, H), one antonym table should be created first.

Example 7:

T16：柏拉图出生于较为富裕的家庭。

H16：柏拉图诞生于清贫家庭。

Antonym Rule: For a given text pair (T, H) which has high similarity, if there exists a pair of antonyms between T and H, it can be justified as textual contradiction.

(6) Negation

The negation is also a good indicator for textual contradiction. The negation "不" in the following Example 8 makes the polarity of T17 and H17 opposite. To calculate the number of negative words in each text, one negation table has been generated. The numbers of the negative words in texts T and H are calculated respectively. If the difference between two numbers is an odd, which indicates the opposite polarity between two texts, the conclusion can be drawn that the text pair is the contradictory one. Negation Rule is created for negative contradiction.

> Example 8:
> T17：草莓不适合运输储存。
> H17：草莓容易运输储存。

Negation Rule: For a given text pair (T, H) which has high similarity, if the difference of the negation numbers of T and H is an odd number, it can be justified as textual contradiction.

4 Experiments

4.1 Experiment Background

The corpus used in the experiments is from NTCIR-10 RITE-2 task containing 814 text pairs for training dataset and 781 text pairs for testing dataset respectively. There are four relation types in the original dataset. In order to aim at contradiction detection, the multiple classification problem is simplified as a binary classification one by marking the other three relation types, forward entailment, bidirectional entailment and independence, as the label "Non-Contradiction".

For individual label, contradiction and non-contradiction, we adopt precision, recall and f1-measure to evaluate as defined as follows.

$$\text{Precision} = \frac{TP}{TP + FP} \tag{2}$$

$$\text{Recall} = \frac{TP}{TP + FN} \tag{3}$$

$$\text{F1 - Measure} = \frac{2 * \text{Precision} * \text{Recall}}{\text{Precision} + \text{Recall}} \tag{4}$$

where TP (True Positives) and FP (False Positives) refer to the numbers of pairs that have been correctly or incorrectly classified as positive respectively. TN (True Negatives) and FN (False Negatives) are the numbers of pairs that have been correctly or incorrectly classified as negative respectively.

In order to verify the efficiency of our approach based on linguistic phenomena, we construct two different experiment systems. The baseline CTCR-nonLP uses statistical features, lexical semantic features and syntactic features to train the SVM classifier and predict the class of the testing Chinese text pairs. CTCR-LP has been introduced contradiction linguistic phenomena and uses semantic rules to recognize contradiction. The experiment results of the two systems are listed as follows and the analysis on them is also provided.

4.2   Experiment Results

CTCR-nonLP combines three kinds of features, including statistical features, lexical semantic features and syntactic features. These features are used to train the classifier and predict the classification result.

The confusion matrix of CTCR-nonLP for NTCIR-10 RITE-2 is listed as follows. The labels in columns are true ones of the text pairs while the labels in rows are given by CTCR-nonLP. According to the confusion matrix in Table 1, we calculate precision, recall and f1-measure for each class as listed in Table 2.

Table 1. Confusion matrix of CTCR-nonLP

Label	Contradiction	Non-Contradiction	
Contradiction	30	11	41
Non-Contradiction	76	664	740
	106	675	781

Table 2. Evaluation results of CTCR-nonLP

Label	Precision	Recall	F1-measure
Contradiction	0.732	0.283	0.408
Non-Contradiction	0.897	0.984	0.938

According to Table 2, in the CTCR-nonLP, the class with the label "Non-Contradiction" has high recall and relatively low precision, which implies the CTCR-nonLP classifies more text pairs in the testing dataset as "Non-Contradiction" while their true class is actually contradiction. Likewise, it's not hard to find this from the confusion matrix in Table 1, where totally 76 text pairs with the label "Contradiction" was classified as "Non-Contradiction". Compared to the label "Non-Contradiction", the accuracy of contradiction recognition is quite low. Only 30 contradictory text pairs have been classified correctly and the recall of the label "Contradiction" is only 0.283, probably due to the insufficient features for the textual contradiction recognition.

After further analysis on the result of CTCR-nonLP, the textual contradiction semantic rules according to linguistic phenomena are used to improve the precision and recall of textual contradiction recognition.

The system CTCR-LP has been added semantic rules according to linguistic phenomena based on CTCR-nonLP. The confusion matrix of CTCR-LP is listed in the following Table 3 and the meaning of the labels is the same as the one in Table 1. According to the confusion matrix in Table 3, we calculate precision, recall and f1-measure for each label as in Table 4.

Table 3. Confusion matrix of CTCR-LP

Label	Contradiction	Non-Contradiction	
Contradiction	58	36	94
Non-Contradiction	48	639	687
	106	675	781

Table 4. Evaluation results of CTCR-LP

Label	Precision	Recall	F1-measure
Contradiction	0.617	0.547	0.580
Non-Contradiction	0.930	0.947	0.938

By comparing Table 3 with Table 1, it can be found that 58 contradictory text pairs in CTCR-LP have been classified into the correct category while only 30 contradictory text pairs in CTCR-nonLP have been recognized correctly, which indicates that contradiction semantic rules according to linguistic phenomena has better ability of recognizing contradictory texts than SVM. However since the semantic rules have confused some labels "Non-Contradiction" with "Contradiction", 36 "Non-Contradictory" text pairs have been recognized as the "Contradiction" ones.

The overall comparative experiment results for the contradiction recognition in the two experiment systems are shown in the following Fig. 2.

From Tables 4 and 2, we can find that contradiction semantic rules improve the recall of contradiction significantly from 0.283 to 0.547. As a result of an "over-recognition" by semantic rules, some "Non-Contradiction" texts have been recognized as "Contradiction", which lead to a reduction of the "Contradiction" precision. Taken precision and recall together, the f1-measure of "Contradiction", which is a comprehensive evaluating indicator of precision and recall, is improved from 0.408 to 0.580. On the whole, CTCR-LP is superior to CTCR-nonLP obviously in terms of contradiction recognition.

In summary, CTCR-nonLP with more statistical features performs poorly on contradiction recognition. CTCR-LP with semantic rules based on linguistic phenomena has improved textual contradiction recognition. The experiment results demonstrate the effectiveness of our textual contradiction recognition approach.

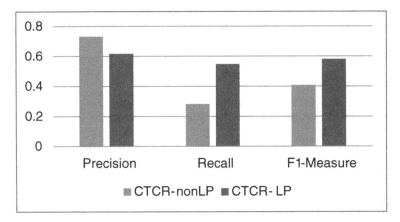

Fig. 2. Experiment results of contradiction recognition in CTCR-nonLP and CTCR-LP

5   Conclusions

In this paper, we have proposed a textual contradiction recognition approach based on semantic rules according to contradictory linguistic phenomena. The experiments on the training dataset and testing dataset provided by NTCIR-10 RITE-2 task has confirmed that our approach has improved Chinese textual contradiction recognition significantly.

The experiment results also show that our contradiction semantic rules have confused some "Non-Contradiction" label with "Contradiction" label in contradiction recognition, which indicates that our semantic rule set is not complete and robust. In future work, we will pay more attentions to deep analysis of linguistic phenomena and the completion of the contradiction semantic rule set to improve the performance of contradiction recognition. Moreover, the combination of contradiction semantic rules should be taken into consideration in the future.

Acknowledgements. The work presented in this paper is supported by the National Natural Science Foundation of China (No. 61100133 and 61173062), the Major Projects of National Social Science Foundation of China (No. 11&ZD189).

References

1. Yuan, Y.L., Wang, M.H.: The inference and identification models for textual entailment. J. Chin. Inf. Process. **24**(2), 3–13 (2010)
2. Wang, K.X.: Semantic Conflict Detection and Its Implementation Based on Implication Reasoning [D]. Shanghai Jiao Tong University, Shanghai (2011)
3. Li, F., Zheng, Z., Tang, Y., Bu, F., Ge, R., Zhu, X., Zhang, X.: Thu quanta at TAC 2008 QA and RTE track. In: Proceedings of the Text Analysis Conference (2008)
4. Liu, M.F., Li, Y., Xiao, Y., Lei, C.Y.: WUST SVM-based system at NTCIR-9 RITE task. In: Proceedings of the 9th NTCIR Conference, pp. 318–324 (2011)

5. Liu, M.F., Wang, Y., Li, Y., Hu, H.J.: WUST at NTCIR-10 RITE-2 task: multiple feature approach to chinese textual entailment. In: Proceedings of the 10th NTCIR Conference (2013)
6. Condoravdi, C., Crouch, D., De Paiva, V., Stolle, R., Bobrow, D.G.: Entailment, intensionality and text understanding. In: Proceedings of the HLT-NAACL 2003 Workshop on Text Meaning, pp. 38–45 (2003)
7. Harabagiu, S., Hickl, A., Lacatusu, F.: Negation, contrast and contradiction in text processing. In: Proceeding of the 21st National Conference on Artificial Intelligence, pp. 755–762 (2006)
8. Ritter, A., Downey, D., Soderland, S., Etzioni, O.: It's a contradiction – no, it's not: a case study using functional relations. In: Proceedings of the Conference on Empirical Methods in Natural Language Processing, pp. 11–20 (2008)
9. De Marneffe, M.C., Rafferty, A.N., Manning, C.D.: Finding contradictions in text. In: Proceeding of the 46th Annual Meeting of the Association for Computational Linguistics: Human Language Technologies, pp. 1039–1047 (2008)
10. Voorhees, E.M.: Contradictions and justifications: extensions to the textual entailment task. In: Proceeding of Association for Computational Linguistics, pp. 63–71 (2008)
11. Shih, C., Lee, C., Tsai, R.T., Hsh, W.: Validating contradiction in texts using online co-mention pattern checking. ACM Trans. Asian Lang. Inf. Process. 11(4), 17 (2012)
12. Schlippe, T., Zhu, C., Lemcke, D., Schultz, T.: Statistical machine translation based text normalization with crowdsourcing. In: 2013 IEEE International Conference on Proceedings of Acoustics, Speech and Signal Processing (ICASSP), IEEE, pp. 8406–8410 (2013)
13. Gao, J., Li, M., Wu, A., Huang, C.N.: Chinese word segmentation and named entity recognition: a pragmatic approach. Comput. Linguist. 31(4), 531–574 (2005)

Event-Oriented Semantic Data Generation for Medical Guidelines

Yuling Fan[1,2(✉)], Jinguang Gu[1,2], and Zhisheng Huang[3]

[1] College of Computer Science and Technology,
Wuhan University of Science and Technology, Wuhan 430065, China
1023293818@qq.com
[2] Hubei Province Key Laboratory of Intelligent Information Processing
and Real-Time Industrial System, Wuhan 430065, China
[3] Departments of Computer Science, VU University Amsterdam,
1081 HV Amsterdam, The Netherlands

Abstract. Clinical practice guidelines aim to help doctors and patients to improve the quality of care, reduce unjustified practice variations and reduce healthcare costs. The study of medical events is of special significance in medicine. However, in most existing medical guidelines, there is a lack of effective methods and tools to describe medical events. This paper analyzes the medical guidelines and the events in those guidelines, discusses the definition of semantic event and the theoretical model of event proposed by Vendler. On this basis, the paper introduces the importance as an additional dimension in the definition on events. Events of guidelines can be extracted according to this definition. We convert those extracted events using the XSLT to generate their RDF semantic data. The generated semantic data are mapped with not only their relevance of the well-known medical ontology such as SNOMED CT, but also used in the system SeSRUA, a semantically-enabled system for rational use of antibiotics. The experiments show that it can promote the rational use of drugs in the development of information technology and knowledge management.

Keywords: Clinical guidelines · Medical events · SNOMED · Semantic web technology

1 Introduction

Clinical practice guidelines (CPGs [1]) are authoritative guidance documents which are varieties of clinical guidance developed by a system that can help doctors and patients to make appropriate treatment, choices, decision for a specific clinical problem [2]. Clinical guidelines knowledge can be divided into consensus-based guidelines and evidence-based guidelines. At present researches on the guidelines are most concentrated on computerized guidelines [2–4] and evidence-based guidelines. In this paper, we propose an approach of semantic data generation for medical guidelines based on events. We will show how the results of semantic data generation of events are able to monitor the actions and observations of care providers and to provide guideline-based advice at the point of care.

© Springer-Verlag Berlin Heidelberg 2014
D. Zhao et al. (Eds.): CSWS 2014, CCIS 480, pp. 123–133, 2014.
DOI: 10.1007/978-3-662-45495-4_11

There exist varieties of definitions on events, which are different from an area to another area. In linguistics, an event is primarily determined by the verb classification, and the property of events decides by its corresponding verb predicate type. As for the expression of the events, Vendler considers the event as one which is constituted by a set of semantic primitives, involving important semantic primitives including actions, time, location etc. Sowa [5] defines an event as a change in the process occurs in discrete steps. In the well-known medical terminologies SNOMED CT, a medical event is defined as a variety of incidents, such as "falling at home", "being hit by a car". There are about thirty thousand medical events listed in SNOMED CT in total. Medical events are apparently different from the events understood in our daily life. In this paper, we take a similar definition with that in SNOMED CT. Namely we define an event in medical guidelines as a special state that should be avoided (i.e., an accident), or a state which may lead to an accident (i.e., an alert state).

Thus, in this paper, we consider the events as the accidents which should be fully avoided or the states which should be alerting. There are different methods to extract those events. In this paper, we use the method of manual preprocessing first, and then convert the guidelines into RDF triples. Using RDF triples to represent events is feasible, on this basis, it can establish the concept mapping between SNOMED CT and the semantic representation of clinical guidelines. We can make querying and reasoning on those semantic data. The semantic representation of clinical guidelines can be used in clinical decision-support systems.

The rest of the paper is organized as follows: Sect. 2 gives a brief description and analysis of the medical guideline. In Sect. 3, it describes the events in medical. Event definitions of the medical guidelines are in the Sect. 4, Sect. 5 describes how to use XSLT to convert events to generate RDF triples, and semantic processing of events. Finally, it summarizes and discusses the work of this paper in Sect. 6.

2　Clinical Guidelines

Trustworthy clinical practice guidelines are documents that include recommendations intended to optimize patient care that are informed by a systematic review of evidence and an assessment of the benefit and harms of alternative care options. Clinical practice guidelines are used to improve the medical process and result and optimize the resource utilization. Those documents include guidelines target, medical background, patient eligibility criteria, evidence, etc. Meanwhile, according to the contents of the different levels of detail, guidelines can be divided into brief summary and complete summary. In [3], it describes in detail the content, structure and characteristics of the guidelines.

So far, the development of guidelines has more than 30 year history. Initially, doctors are reluctant to use the guidelines because of the patient's individual differences. With the advent of decision support systems, guidelines are gradually formalized and computerized, the status of guidelines in clinical is more important. Therefore, how to represent computer-interpretable guidelines has become an important research topic in the field of medical guidelines.

Early guidelines are based on the consensus of doctors and specialists. With the development of guidelines, more and more guidelines are evidence-based. Thus, guidelines knowledge can be divided into consensus-based guideline and evidence-based guideline. Evidence-based guidelines emphasize evidence, which requires researchers to provide the most reliable clinical research evidence as possible, and knowledge of evidence-based guidelines are more authoritative, since it is validated and recommend by a collective of experts in the field. Research on evidence-based clinical guidelines for physicians and other medical staffs in the medical diagnosis provide more accurate proposals, and can clearly define the medical procedures and reduce medical disputes. While in this paper, the study on the guidelines of events is consensus-based clinical guidelines.

3   Events in Medical

3.1   Semantic Event

From the perspective of the event semantics, verb is a collection of events and the classification of verb determines the nature of the various event structures. Linguist Vendler divides verbs into four categories according to the time characteristics of the verbs: state verbs, active verbs, reached verbs and completed verbs. The characteristics of state verbs are no starting and ending, the nature of the event described in the different stages of the continuous condition is the same, such as have, know. Active verbs, also known as process verbs, have a starting point, but no ending point, describe the process with homogeneous characteristics, such as run, walk, etc. Reached verbs are instantaneous verbs, which mean it is done instantaneously and has no process, like find, reach. Completed verbs have starting point and termination point, such as build a house, make a chair and so on.

Corresponding to the above four categories of the verbs, events can be divided into state events, active events, reached events and completed events. The first three are the basic event types, and completed event is an event with the combination of active event and reached event. For example, "John painted the house" is an active event, while "John painted the house red" is a completed event, it's obvious that completed event is a combination of the basic types of events. Various semantic relations between events and internal semantic structure of events can be represented by a diagram. "Lisa is going to watch a movie on Sunday." For such sentence, available the following diagram (Fig. 1):

A, P, T and <fut> in the picture above separately refer to agent, patient, time and the future. The first three are semantic roles of events, the last is an event semantic operator, event predicate of atomic time is verb "watch".

Vendler considers the event as one which is constituted by a set of semantic primitives, involving important semantic primitives including actions, time, location, etc. [6, 7]. The theoretical model of event proposed by Vendler has guiding significance for the subsequent construction work of event model. Consider the example "Jones buttered the toast", the corresponding logical expression is: BUTTER (JONES, THE TOAST). But you extend this sentence as "Jones buttered the toast somewhere with something at some time". The BUTTER can have a different number of arguments. Davidson [8] proposes a solution that BUTTER is still seen as a predicate with a fixed

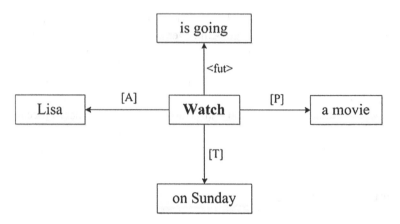

Fig. 1. Atomic events figure of watch

number of arguments, using conjunctive relations to reflect additional information. Adding a new argument of event "e", such expression depicts the semantic differences more accurately.

In the following two sentences (a) and (c), (b) and (d) are the expression using the Davidson method:

(a) Jones buttered the toast in the bathroom.
(b) ($^\exists$e (BUTTER (JONES, THE TOAST, e) IN (THE BATHROOM, e)).
(c) Jones buttered the toast near the bathroom.
(d) ($^\exists$e) (BUTTER (JONES, THE TOAST, e) NEAR (THE BATHROOM, e)).

As you can see from the above example, adding a new event argument "e" can reflect the differences between semantics in more detail.

3.2   Medical Events

Converting large number of discrete natural language information into structured information that computer can handle, one of the most important prerequisite is to define the relevant data model. According to the analysis of background knowledge in the medical field as well as the expansion of the traditional event model, in this paper, we define an event model.

Based on the study of semantic primitives, we think event is constituted by a set of semantic primitives. Important semantic primitives involving in medical events include actions, time, location, etc. That is Event = (object, time, location).

If medical events consider to be a class M, then diagnostic events and medication events are subclasses of M, which can be represented as M = {E1 ∪ E2 ∪ E3. . . ∪ En}. Each subclass Ex contains examples of events E that belong to the category, and are constituted by event elements, which is E = <object, time, location>.

4   Events in Medical Guidelines

There are different definitions of the events in the literature. Some literature defines the event as an action. The common understanding of the action is that an action can cause the change of state, which varies from one state to another. If you are required to enumerate the events in your life, no one will list the change of states, but cite the important things which have a significant impact. Based on this, this paper add the importance as an additional dimension in the definition of the event, that events are more important things which will have a great impact on life.

In this paper, we consider the events as the accidents which should be fully avoided or the states which should be alerting. Guidelines are a collection of diagnosis, when facing a state, what is absolutely prohibited, what must be done, which is known as event correlation. The importance of guidelines is not subjective viewpoint, in this paper, the definition of events reflected in medical guidelines is "precautions". If state A and state B simultaneously is absolutely prohibited, the state is an event. But state A and B separately is not necessarily an event. If state A, then should execute state B, for such a sentence, A is an event but B is not an important state. Thus, in the precautions of guidelines, not all states are events.

This paper takes guidelines for rational use of antibiotics as test data. The description of β-lactams/β-Lactamaseinhibitors in the guidelines is as follows: "When using this drug, if allergic reaction occurs, you should be discontinued immediately; Once the anaphylactic shock occurs, you should local rescue, and give oxygen and an injection of epinephrine, adrenocorticotropic hormone and other anti-shock treatment" (应用本类药物时如发生过敏反应, 须立即停药; 一旦发生过敏性休克, 应就地抢救, 并给予吸氧及注射肾上腺素、肾上腺皮质激素等抗休克治疗). State A is "using β-lactams/β-Lactamaseinhibitors", state B is "allergic reaction", A + B is a state that should be avoided, that is, "stopping the medication immediately" is an event. Similarly, C is "anaphylactic shock", then A + C is also a state that should be avoided, so "local rescue" and "give oxygen" and "an injection of epinephrine, adrenocortico-tropic hormone" and "other anti-shock treatments" are events.

Actually, for "When using this drug, if allergic reaction occurs, you should be discontinued immediately (应用本类药物时如发生过敏反应, 须立即停药)", stopping the medication immediately is an event, and it's event precondition is "When using this drug, if allergic reaction occurs", "should" is the event operator.

5   Semantic Data Generation and Processing of Events

In this paper, we propose an approach of semantic data generation for medical guidelines based on events. To convert events to generate their RDF semantic data, what we need to do first is to convert textual guidelines into computer-interpretable guidelines. So far, there exist varieties of methods on events extraction, like sentence segmentation, POS tagging and term annotation [9], machine learning method [10], pattern matching method [11] and rule-based event extraction method [12]. But considering the accurateness of this study, we extract events of guidelines in this paper using manual preprocess. There are two reasons for it. One is that the total amount of guideline statements (e.g., usually less than a few hundred ones) is not too much. Its manual processing would not take too much time; The second reason is that we can use natural language processing tools for a help to extract events.

In this paper, we use the method of manual preprocessing first, and then convert the XML documents into RDF triples by using XSLT. It also establishes the concept mapping between SNOMED CT and guidelines, so we can make querying and reasoning on those semantic representation of clinical guidelines.

5.1   Guidelines Preprocess

In order to ensure the accurateness of the study, when manually extract the events, we pay attention to the following two points: One is that every statement must have an event precondition, an event operator, and an event; The second is abbreviated words in the guidelines like "should (须)", "may (可)" must be extended as "should (必须)", "may (可以)".

This paper takes the guidelines for rational use of antibiotics which is published by the Chinese health authorities as the test data. For example, the description of Penicillin antibiotics in the guidelines is as follows: "Once the anaphylactic shock occurs, you should local rescue, and give oxygen and an injection of epinephrine, adrenocorticotropic hormone and other anti-shock treatment (过敏性休克一旦发生, 必须就地抢救, 并立即给病人注射肾上腺素, 并给予吸氧、应用升压药、肾上腺皮质激素等抗休克治疗)". After manual preprocessing, the part of the resulting document is generated as follows (Fig. 2):

```
<statements>
    <text>过敏性休克一旦发生，必须就地抢救，并立即给病人注射肾上腺素，
    并给予吸氧、应用升压药、肾上腺皮质激素等抗休克治疗。</text>
    <statement>
        <eventPrecondition>青霉素类抗生素过敏性休克一旦发生</eventPrecondition>
        <eventOperator>必须</eventOperator>
        <event>就地抢救</event>
    </statement>
    <statement>
        <eventPrecondition>青霉素类抗生素过敏性休克一旦发生</eventPrecondition>
        <eventOperator>必须</eventOperator>
        <event>立即给病人注射肾上腺素</event>
    </statement>
    <statement>
        <eventPrecondition>青霉素类抗生素过敏性休克一旦发生</eventPrecondition>
        <eventOperator>必须</eventOperator>
        <event>给予吸氧、应用升压药、肾上腺皮质激素等抗休克治疗</event>
    </statement>
</statements>
```

Fig. 2. Result of event extraction manually

5.2   Generation of Semantic Data

XSLT (Extensible Stylesheet Language Transformations) [13] is an extensible stylesheet language. This paper uses XSLT to transform XML documents into RDF triples. XSLT uses XPath [14] to find information in a XML document, and XPath is used to navigate the XML documents, which rely on the elements and attributes in the XML documents. During the process of conversion, XSLT uses XPath to define source documents that can be matched to one or more predefined templates, once a match is found, XSLT will convert the matching part of the source document into the result document.

The final format of conversion is RDF triples. RDF can not only describe the resources by attributes and the attributes values, but also the relationship between resources. RDF can be represented using a directed graph, each edge of the graph corresponds to a "subject-predicate-object" triples. In fact, each statement can be expressed as RDF triples <subject, predicate, object>, that is, the predicate, the subject, and the object of the statement.

This paper uses an XSL document to convert the xml document into a RDF Ntriple file. The part of the XSL document is shown in Fig. 3.

```
<xsl:for-each select="statements">

 <xsl:variable name="statementsid" select="concat($id,'_', position())"/>

 <xsl:text>&lt;</xsl:text><xsl:value-of select="$id"/><xsl:text>&gt;&lt;</xsl:text>
 <xsl:value-of select="$sct"/><xsl:text>hasStatements&gt;&lt;</xsl:text>
 <xsl:value-of select="$statementsid"/><xsl:text>&gt; .
 </xsl:text>

 <xsl:call-template name="statements">
 <xsl:with-param name="statementsid" select="$statementsid"/>
 </xsl:call-template>


 </xsl:for-each>
```

Fig. 3. XSL document

The final format is RDF Ntriples, and the part of RDF triples is shown as follows:

```
<http://wasp.cs.vu.nl/sct/id#gl001zsh140331><http://wasp.cs.vu.nl/sct/sct#hasStatements><http://wasp.cs.vu.nl/sct/id#gl001zsh140331_2> .
<http://wasp.cs.vu.nl/sct/id#gl001zsh140331_2><http://wasp.cs.vu.nl/sct/sct#Text>"过敏性休克一旦发生，必须就地抢救，并立即给病人注射肾上腺素，
<http://wasp.cs.vu.nl/sct/id#gl001zsh140331_2><http://wasp.cs.vu.nl/sct/sct#hasStatement><http://wasp.cs.vu.nl/sct/id#gl001zsh140331_21>
<http://wasp.cs.vu.nl/sct/id#gl001zsh140331_21><http://wasp.cs.vu.nl/sct/sct#haseventPrecondition>"青霉素类抗生素过敏性休克一旦发生"
<http://wasp.cs.vu.nl/sct/id#gl001zsh140331_21><http://wasp.cs.vu.nl/sct/sct#haseventOperator><http://wasp.cs.vu.nl/sct/guideline#Must>
  <http://wasp.cs.vu.nl/sct/guideline#Must><http://www.w3.org/2000/01/rdf-schema#label>"必须"@cn.
  <http://wasp.cs.vu.nl/sct/guideline#Must><http://www.w3.org/2000/01/rdf-schema#label>"Must"@en.
  <http://wasp.cs.vu.nl/sct/id#gl001zsh140331_21><http://wasp.cs.vu.nl/sct/sct#hasevent >"就地抢救"
<http://wasp.cs.vu.nl/sct/id#gl001zsh140331_2><http://wasp.cs.vu.nl/sct/sct#hasStatement><http://wasp.cs.vu.nl/sct/id#gl001zsh140331_22>
<http://wasp.cs.vu.nl/sct/id#gl001zsh140331_22><http://wasp.cs.vu.nl/sct/sct#haseventPrecondition>"青霉素类抗生素过敏性休克一旦发生"
<http://wasp.cs.vu.nl/sct/id#gl001zsh140331_22><http://wasp.cs.vu.nl/sct/sct#haseventOperator><http://wasp.cs.vu.nl/sct/guideline#Must>
  <http://wasp.cs.vu.nl/sct/guideline#Must><http://www.w3.org/2000/01/rdf-schema#label>"必须"@cn.
  <http://wasp.cs.vu.nl/sct/guideline#Must><http://www.w3.org/2000/01/rdf-schema#label>"Must"@en.
  <http://wasp.cs.vu.nl/sct/id#gl001zsh140331_22><http://wasp.cs.vu.nl/sct/sct#hasevent >"给病人注射肾上腺素"
<http://wasp.cs.vu.nl/sct/id#gl001zsh140331_2><http://wasp.cs.vu.nl/sct/sct#hasStatement><http://wasp.cs.vu.nl/sct/id#gl001zsh140331_23>
<http://wasp.cs.vu.nl/sct/id#gl001zsh140331_23><http://wasp.cs.vu.nl/sct/sct#haseventPrecondition>"青霉素类抗生素过敏性休克一旦发生"
<http://wasp.cs.vu.nl/sct/id#gl001zsh140331_23><http://wasp.cs.vu.nl/sct/sct#haseventOperator><http://wasp.cs.vu.nl/sct/guideline#Must>
  <http://wasp.cs.vu.nl/sct/guideline#Must><http://www.w3.org/2000/01/rdf-schema#label>"必须"@cn.
  <http://wasp.cs.vu.nl/sct/guideline#Must><http://www.w3.org/2000/01/rdf-schema#label>"Must"@en.
  <http://wasp.cs.vu.nl/sct/id#gl001zsh140331_23><http://wasp.cs.vu.nl/sct/sct#hasevent >"给予吸氧、应用升压药、肾上腺皮质激素等抗休克治疗"
```

Fig. 4. RDF triples

Figure 4 shows that the event operators in triples like "Must (必须)" should be represented in English, like this:

<http://wasp.cs.vu.nl/sct/guideline#Must>;   <http://www.w3.org/2000/01/rdf-schema#label>; "Must" @en;

The "Must" which is linked to "http://wasp.cs.vu.nl/sct/guideline", which establishes their relevance with the well-known medical ontology.

From the analysis above, we can see that the structured degree of semantic data corresponding to such an event is not very high, such as "give oxygen and use vasopressor drugs, adrenocorticotropic hormone and other anti-shock treatment (给予吸氧、应用升压药、肾上腺皮质激素等抗休克治疗)". The entire sentence is simply represented as a description of a single event. But in fact, it is still needed to be cut into more structured parts, such as "give oxygen", "use vasopressor drugs", "adrenocorticotropic hormone" and "anti-shock treatment", which requires us to carry out further semantic processing on this basis. An effective way to achieve it is to establish the concept mapping between SNOMED CT and the semantic representation of clinical guidelines. More precisely, it is to establish their corresponding semantic annotations.

5.3   Semantic Processing

Based on the semantic data generation of events, we can create their relevance with the well-known medical ontology such as SNOMED CT, which can be done by the medical professional dictionary in both English and Chinese and SNOMED CT online server. Firstly, we use the natural language processing tools to extract medical concepts from Chinese documents, we parse XML documents and custom vocabularies. Finally, we use the medical terms RESTAPI tools in both Chinese and English which is based on the LarKC [15] platform to convert the Chinese medical terms into English SNOMED CT.

In this paper, we use the Dom4j package to parse "guideline.xml" for translation of medical concepts into the Chinese documents. Figure 5 is a result of the translation:

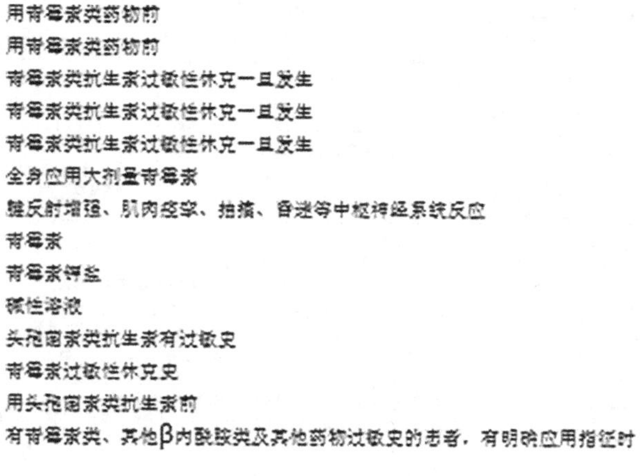

Fig. 5. Result of XML parsing

For simple sentence parsed, keywords extraction involves symptoms, disease, medicines and medical terms, etc. The word segmentation tool we use in this paper is ICTCLAS (Institute of Computing Technology, Chinese Lexical Analysis System) [16] which come from the Chinese academy of sciences. The main characteristics of ICTCLAS include: Chinese word segmentation, POS tagging, named entity recognition, etc. At the same time, it also supports custom vocabularies, which is greatly helpful to extract keywords and phrases in guidelines like symptoms, disease, medicines and medical terms, etc.

We define the vocabularies when using ICTCLAS, for example, "penicillin" can specify part of speech of "YP", namely drug. We can also specify the "fever" as "ZZ", namely symptoms. The vocabularies we define are a large library, which contains all kinds of drugs, symptoms, disease, qualitative values and other medical terms. The complement of the custom vocabularies determines the accuracy of keywords extraction (Fig. 6).

```
用/p 青霉素类/YP 药物/n 前/f
用/p 青霉素类/YP 药物/n 前/f
青霉素类抗生素/YP 过敏性休克/ZZ 一旦/d 发生/v
青霉素类抗生素/YP 过敏性休克/ZZ 一旦/d 发生/v
青霉素类抗生素/YP 过敏性休克/ZZ 一旦/d 发生/v
全身/n 应用/v 大/a 剂量/n 青霉素/n
恶反射增强/ZZ 、/wn 肌肉痉挛/ZZ 、/wn 抽搐/ZZ 、/wn 昏迷/ZZ 等/v 中枢神经系统/ZZ 反应/vi
青霉素/n
青霉素钾盐/YP
```

Fig. 6. Result of POS tagging

After extracting Chinese concepts, we use the medical terms RESTAPI tools in both Chinese and English to convert the Chinese medical terms into English SNOMED CT. The corresponding translation of labels using the REST parameter to turn a language (form) of a term (term) to another language (to) is as follows, medical term "penicillin (青霉素)" as an example:

Command: commandType = getConceptTranslation
Term: conceptterm = 青霉素
Current language: from = cn/en
Target language: to = en/cn
Exact match: exactmatch = yes/no
Result type: resulttype = json/xml
Result limit: limit = 100

The result of the concept mapping between SNOMED CT and the semantic representation of clinical guidelines, "give oxygen and use vasopressor drugs, adrenocorticotropic hormone and other anti-shock treatment (给予吸氧、应用升压药、肾上腺皮质激素等抗休克治疗)" is shown in Fig. 7:

```
<http://wasp.cs.vu.nl/sct/id#ql001zsh140331_23><http://wasp.cs.vu.nl/sct/sct#hasevent >"给予吸氧、应用升压药、""@en.等抗休克治疗".
<http://wasp.cs.vu.nl/sct/id#ql001zsh140331_23>< http://wasp.cs.vu.nl/sct/sct#haseventID> <http://wasp.cs.vu.nl/sct/id#ql001zsh140331_23_1>.
<http://wasp.cs.vu.nl/sct/id#ql001zsh140331_23_1><http://wasp.cs.vu.nl/sct/sct#hasConcept> <http://www.ihtsdo.org/SCT_57485005>.
<http://www.ihtsdo.org/SCT_57485005><http://www.w3.org/2000/01/rdf-schema#label>"Oxygen therapy (procedure)"@en.
<http://www.ihtsdo.org/SCT_57485005><http://www.w3.org/2000/01/rdf-schema#label>"吸氧治疗 (过程)"@cn.
<http://wasp.cs.vu.nl/sct/id#ql001zsh140331_23_1><http://wasp.cs.vu.nl/sct/sct#hasConcept> <http://www.ihtsdo.org/SCT_40789008>.
<http://www.ihtsdo.org/SCT_40789008><http://www.w3.org/2000/01/rdf-schema#label>"Adrenocorticotropic hormone (substance)"@en.
<http://www.ihtsdo.org/SCT_40789008><http://www.w3.org/2000/01/rdf-schema#label>" 肾上腺皮质激素 (物质)"@cn.
```

Fig. 7. Result of relevance with SNOMED CT

Here, we map the event to its SNOMED concepts, and mark the concepts with the corresponding Chinese and English. Thus, we can use a semantic data processing platform, such as LarKC system to process them.

6   Discussion and Conclusion

In this paper, we propose an approach of semantic data generation for medical guidelines based on events. We takes the guidelines for rational use of antibiotics which is published by the Chinese health authorities as the test data. We already have a preliminary result by using the method proposed above. The introduction of the data of events makes it possible to establish a basic set of events for the use of antibiotic agents. These sets of events constitute a collection of states that diagnosis system and clinical decision-support systems should pay particular attention. We are able to use this particular set of states to monitor the actions and observations of care providers and to provide guideline-based advice at the point of care.

However, there are still a lot of future work to do in this paper:

1. Automatic or semi-automatic preprocess. In order to ensure the accurateness of the study, we extract events of guidelines in this paper using manual preprocess instead of natural language tools. In the further study, we will focus on the method of automatic or semi-automatic to extract events.
2. More complete keywords extraction. Keywords extraction mainly depends on the custom vocabularies. Gradually complete vocabularies will make keywords extraction more accurate.
3. Improvement of the correctness in the conversion of concepts in both English and Chinese. Chinese medical terms may not find the corresponding concepts on the SNOMED. In this case, we can only translate them into corresponding English with the help of some dictionaries. But with the complement of SNOMED, Chinese keywords extraction will completely depend on the SNOMED.

References

1. Field, M.J., Lohr, K.N. (eds.): Guidelines for Clinical Practice: from Development to Use. National Academy Press, Washington, D.C. (1992)
2. Zhao, Y., Cui, S.: The research of international clinical practice guidelines. Clin. Educ. General Pract. **2**(3), 176–178 (2004). (赵亚利, 崔树起. 国际临床实践指南的研究进展. 全科医学临床与教育 **2**(3), 176–178 (2004))
3. He, Y., Sun, H.: The research of computerized clinical practice guidelines. Chin. Digit. Med. **2**(1), 10–15 (2007). (何雨生, 孙宏宇. 计算机化临床实践指南研究进展. 中国数字医学 2 (1), 10–15 (2007))
4. Li, Y., Zhao, J., Li, W.: The research and implement of computerized clinical practice guidelines. Chin. J. Med. Instrum. **33**(6), 407–409 (2009). (李毅, 赵军平, 李韦章. 计算机化临床实践指南的研究和实现. 中国医疗器械杂志 **33**(6), 407–409 (2009))
5. Sowa, J.F.: Knowledge Representation: Logical, Philosophical, and Computational Foundations. Brooks/Cole, Pacific Grove (1999)
6. Wu, P.: Formal semantic analysis of completion events. Foreign Lang. Teach. **4**, 41–45 (2010). (吴平. 完成事件的形式语义分析. 外语与外语教学 **4**, 41–45 (2010))
7. Wu, P.: Formal semantic analysis of completion events. Foreign Lang. Teach. **4**, 8–12 (2007). (吴平. 试论事件语义学的研究方法. 外语与外语教学 **4**, 8–12 (2007))
8. Davidson, D.: The logical form of action sentences. In: Researcher, N. (ed.) The Logic of Decision and Action, pp. 81–120. University of Pittsburgh Press, Pittsburgh (1967)
9. Kaljuand K, Schneider G, Rinaldi F. UZurich in the BioNLP 2009 Shared Task. I //BioNLP '09 Proc of the Workshop on BioNLP: Shared Task, Stroudsburg, PA, USA: Association for Computational Linguistics, 2009:28–36
10. Bjorne, J., Heimonen, J., Ginter, F., et al.: Extracting complex biological events with rich graph-based feature sets. In: BioNLP '09 Proceedings of the Workshop on BioNLP: Shared Task, pp. 10–18 Association for Computational Linguistics, Stroudsburg (2009)
11. Bretonnel Cohen, K., Verspoor, K., Johnson, H.L., et al.: High-precision biological event extraction with a concept recognizer. In: BioNLP '09 Proceedings of the Workshop on BioNLP: Shared Task, pp. 50–58. Association for Computational Linguistics, Stroudsburg (2009)
12. Sarafraz, F., Eales, J., Mohammadi, R., et al.: Biomedical event detection using rules, conditional random fields and parse tree distances. In: BioNLP '09 Proceedings of the Workshop on BioNLP: Shared Task, pp. 115–118. Association for Computational Linguistics, Stroudsburg (2009)
13. XSL Transformations (XSLT) [EB/OL], 18 August 1998. http://www.w3.org/TR/xslt. Accessed 12 May 2014
14. XML Path Language (XPath) [EB/OL], 21 April 1999. http://www.w3.org/TR/xpath/
15. Large Knowledge Collider [EB/OL], 1 April 2008. http://www.larkc.eu. Accessed 8 May 2014
16. Zhang, H., Yu, H., Xiong, D., Liu, Q.: HHMM-based Chinese lexical analyzer ICTCLAS. In: SIGHAN '03 Proceedings of the Second SIGHAN Workshop on Chinese Language Processing, pp. 184–187 (2003)

Ontology Evolution Detection: Method and Results

Gaofan Li[1,2], Peng Wang[1,2(✉)], and Bin Yu[3]

[1] School of Computer Science and Engineering,
Southeast University, Nanjing, China
lgvancl@gmail.com, pwang@seu.edu.cn
[2] College of Software Engineering, Southeast University, Nanjing, China
[3] Communication Station of Unit 95028, P.L.A., Wuhan, China

Abstract. The distributed and dynamic characteristics of semantic Web make ontologies change constantly, that leads to ontology evolution and different versions of ontologies. In ontology evolution, changes would happen at domain, shared conceptual model and ontology representation, and then these changes would propagate and affect semantics of other elements. This paper focuses on ontology evolution detection, which is the foundation of ontology evolution problem. First, an evolution detection method based on similarity is proposed. Then the minimum cost of evolution strategy is also proposed to recover the reasonable evolution process. Experimental results show that this method performs well on OAEI benchmark data set.

Keywords: Ontology · Ontology evolution · Evolution detection

1 Motivation

Distributed and dynamic characteristics of the Web make ontologies change frequently, which lead to the ontology evolution and derive different versions of ontologies. When an ontology evolves, it will affect all applications based on this ontology. For example, existing query or reason results would be wrong, and matching results with other ontologies need to be calculated again. Therefore, understanding ontology evolution is a key task for ontology management and developing practical semantic Web applications. Ontology evolution can be divided to six phases: capturing, representation, semantic of change, implementation, propagation, validation [1]. Therefore, ontology evolution detection, namely, capturing changes in ontology evolution, is the basis of handling the ontology evolution problem [2–4]. In addition, it is necessary to know how such evolutions happen, that is useful to repair the errors brought by ontology evolution with low cost. For instance, we can only update few reasoning results affected by ontology evolution.

For ontology evolution detection, most works handle the change operations with different levels. Some of them propose the evolution structures to find the changes [8, 9]. Others present some algorithms. For example, Tury and Bieliková summarize three ways to find changes: (1) searching the same elements in ontology; (2) making use of heuristic algorithm; (3) using intelligent algorithm such as neural network.

© Springer-Verlag Berlin Heidelberg 2014
D. Zhao et al. (Eds.): CSWS 2014, CCIS 480, pp. 134–145, 2014.
DOI: 10.1007/978-3-662-45495-4_12

Hartung et al. use an *diff* algorithm (available in the CODEX web tool to identify the changes that occurred between two versions of an ontology [12, 15]. The algorithm uses a set of rules to first identify basic changes (add/update/delete) that are then aggregated into a smaller set of more complex (semantic) changes, such as merge, split or changes of entire subgraphs.

After detecting the evolution of the ontology, it needs to further describe the process of the ontology evolution. For this problem, some works usually integrated it into the evolution structure and they focused on the evolution propagation. Palma et al. proposed the change management model and strategy of ontology in distributed environment [10]. Stojanovic et al. defined some elementary evolution strategies and proposed some advanced evolutionary strategies on that basis: structure-driven strategy, process-driven strategy, instance-driven strategy, frequency-driven strategy [11].

Once the ontology changes, experts should do the job of mapping for the new version. The work would be very complex and time-consuming if the ontology was very large. And if the scope of impact of the evolution could be determined quantitatively and accurately, thus only local of the ontology should be re-mapping, which would reduce the cost of re-mapping. Hartung et al. proposed two measurement of the stability [12]. Lee et al. import impact factor and use measurement of complexity to measure the evolution [13].

This paper focuses on ontology evolution detection, the evolution strategy with minimum cost and the impact of the evolution. A practical similarity-based ontology evolution detection method is proposed. Then the algorithm of finding the minimum cost strategy of ontology evolution is discussed. And we give three ideas to measure the evolution impact. Finally, we demonstrate the effectiveness on OAEI benchmark data set.

2 Method

2.1 Evolution Detection

Text is one of the main carrier and communication medium of human language. Ontology creators usually use general vocabularies to label or annotate ontology elements. When an ontology evolves, ontology creators would change some text information or semantic information of elements.

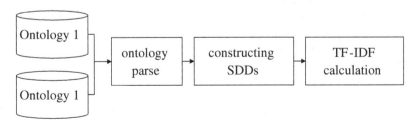

Fig. 1. Similarity-based evolution detection method

Therefore, there must be a method to capture both text and semantic changes. This paper proposes a similarity-based evolution detection method, and it employs semantic description document (SDD) to detect both text and semantic changes. Flow chart of the method is shown in Fig. 1.

For an ontology element, this paper organizes related information based on the semantic description. We call this organization way the semantic description document (SDD) [5]. To avoid introducing unrelated text, SDD is constrained in the semantic subgraph of an element. In additional, SDD does not consider ontology language metadata, such as *rdfs:Class* and *owl:hasValue*. During SDD construction, the text preprocessing contains stemming and removing frequent vocabularies. Process of constructing SDDs is shown in Fig. 2.

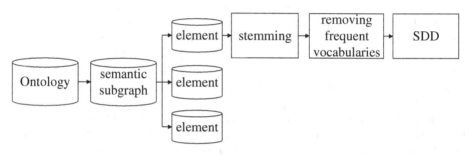

Fig. 2. SDD construction

For each concept, relation or instance, it has a basic SDD, which is consisted by local name, label and annotation. The basic SDD is:

$$DS_{base}(e) = \alpha_1{}^*W_{localname} + \alpha_2{}^*W_{label} + \alpha_3{}^*W_{comment} + \alpha_4{}^*W_{otherAnnotation}$$

where $W_{localname}$ is local name, W_{label} is *rdfs:label* text, $W_{comment}$ is the *rdfs:comment* text, and $W_{otherAnnotation}$ is other annotation text. Weight α_i for each kind of text is in [0, 1]. Therefore, SDD is the set of words with weights and denotes the union operation between sets.

Besides the basic SDD, other SDDs describe concept hierarchy, property hierarchy, property domain and range, et al. Generally, we consider the SDD in two sides: (1) SDD can re-organize text information according to the semantic description of elements; (2) To avoid SDD containing irrelevant and unimportant text information, SDD is constrained in semantic subgraphs [5].

After constructing SDD for concepts and properties, for given two ontologies with different versions, 1-1 alignments can be discovered by computing similarities between SDD. A SDD is a set of vocabularies with weights, namely, $DS = \{p_1{}^*W_1, p_2{}^*W_2, p_x{}^*W_x\}$. We can use cosine to measure the similarities. Let $Doc = \{DS_1, DS_2, DS_N\}$ be the set of same type SDD, and it contains n items $t_1, t_2, \ldots, t_n$. Thus each document DS_i can be described as an n-dimension vector $D_i = (d_{i1}, d_{i2}, :::, d_{in})$, where d_{ij} is the weight of j-th item. The weight d_{ij} in vector D_i is TF-IDF weight. The similarity between two

SDDs is the cosine value of their corresponding vectors. Therefore, the similarity between D_i and D_j is:

$$Sim(D_i, D_j) = \frac{\sum_{k=1}^{n} d_{ik} \times d_{jk}}{\sqrt{\sum_{k=1}^{n} d_{ik}^2 \times \sum_{k=1}^{n} d_{jk}^2}}$$

In our previous work [6, 7], we demonstrate that the similarity based on SDD is very effective for finding alignments between ontologies. However, the goal of evolution detection is finding differences or changes between ontologies; this is different to ontology alignments. First, all unmatched elements are treated as changed in ontology evolution. Then we set a threshold 0.90 for 1-1 alignments. If an alignment is greater than 0.90, the corresponding elements have no changes. On the contrary, elements have changed. Thus the evolution can be detected.

2.2   Evolution Strategy with Minimum Cost

After we detect the evolution of the ontology, we want to further describe the process of the ontology evolution. This paper proposes an algorithm to find the evolution strategy with minimum cost. But there are some problems to be solved before the algorithm:

- How to handle the operations.
- The ontology is in an uncertain state after performing an operation in evolution (decision point), how to handle this situation.
- How to choose the change operations with less impact.
- How to measure impact.

For the first problem, we divide change operations into two levels: basic change operations and composite change operations as shown in Fig. 3. Basic change operations only modify one element and cannot be divided into other operations. Composite change operations modify multiple entities or modify only one entity many times.

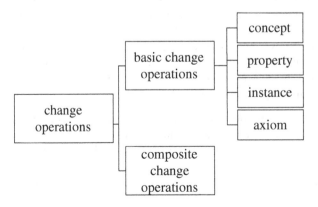

Fig. 3. Change operations

After performing an operation in evolution, the ontology is in an uncertain state, and we call this stage the decision point.

Then we propose the conception of cost to measure impact. For each change operations Ch we defined has the value of the cost C_{Ch}, which determines the degree of impact on ontology. Evolution strategy contains a series of change operations, so the total cost of the evolution strategy C_{st} is:

$$C_{st} = \sum_{i=1}^{j} n_{ch_i} \times C_{ch_i}$$

Therefore, this paper proposed an algorithm to find the evolution strategy with minimum cost. The problem is how to choose the change operations to complete the evolution strategy.

As we know, the ontology is in an uncertain state after executing the decision point, that is to say there may be several evolution paths (change operations) at each decision point. Given two continuous ontology versions, there are many possible paths consisting of operations. We call a path an evolution strategy.

So we define a series of heuristic rules following behind each decision point which include most possible paths. The rules may be basic or composite change operations and it could also be another decision point. Therefore, executing the decision point could bring about derivative actions. One of the goals of the algorithm is executing at the decision point recursively until there are not any derivative actions.

Each evolution path (strategy) may have different value of cost because of different choice of change operations. Obviously the algorithm would choose the evolution strategy with minimum cost.

The algorithm which is to find the evolution strategy with minimum cost is as Algorithm 1 shows. Let RList be operation set, DList be decision set, RulesList be decision rules. For each change operation Ch, we defined its cost CCh. Evolution strategy contains a series of change operations. The evolution strategy with minimum cost is to find a series of change operations which have the minimum cost. At the beginning of the algorithm, we check the operation Ch. If it is not in DList, which means there are not any derivative actions after Ch, we execute the Ch directly and put it into EvoList. Otherwise we execute each rule after this decision point until there is no any derivative action. That would form a series of paths and then we calculate cost of each path, and choose the one with minimum cost as the final path. We put all operations in the path into EvoList, which is the evolution strategy with minimum cost.

In conclusion the algorithm which is to find the evolution strategy with minimum cost can be summed up as follows:

- Executing the decision point could bring about derivative actions.
- We proposed some heuristic rules (change operations) after each decision point.
- We execute at the decision point recursively until there are not any derivative actions.
- We calculate cost of each path and choose the evolution strategy with minimum cost.

What calls for special attention is that the value of the impact of each change operation is decided by experience and the structure of the ontology. In practice the value should be discussed by both ontology engineers and domain experts, and varies from the different framework of ontology structure.

Algorithm 1. Minimum Cost Strategy of Ontology Evolution	
Input: ontology version of O, external request list *RList*, decision point list *DList*, Decision rules list *RulesList*	
Out: Evolution list *EvoList*, using defined change operations	
1	O_s:current ontology state
2	**Begin**
3	**foreach** *RList*
4	**If** $R_n = Ch_n$ is not in *DList* **then**
5	$Ch_n = Evo_n$
6	$O_s = Ch_n(O)$
7	**Return** *RList*
8	**else**
9	$O_s = Ch_n(O_s)$
10	**foreach** *Ru* in *RulesList* **do**
11	Decision（d）Executive corresponding decision strategies
12	ComputeCast（O_s, d）compute cost of each strategy
13	**If** there is operation derived from strategy **then**
14	Decision（d'）Executive corresponding decision strategies
15	ComputeCast（O_s, d'）compute cost of each strategy in new decision point
16	**End**
17	**End**
18	Choose minimum cost strategy path of Ch_n, $Evo_n = (Ch_n, Ch_{n1}, Ch_{n2}...)$
19	Put all corresponding operations into *EvoList*
20	**End**

2.3   Evolution Impact Analysis

After obtaining the specific information of ontology evolution, we want to further analyze the impact brought by evolution. Evolution impact analysis includes measuring the impact quantitatively and getting the scope of the ontology evolution. This paper proposed three ideas to measure the evolution impact.

For our measures, we use the following cardinalities for elements of the different ontology versions.

$|E_i|$, $|E_j|$ is the number of elements in different ontology versions (O_i and O_j).
$|E_i \cap E_j|$ is the number of overlapping elements between O_i and O_j.
$|E_i/E_j|$ is the number of elements only in O_i but not in O_j.
$|E_j/E_i|$ is the number of elements only in O_j but not in O_i.

Then the *Basic stability* can be calculated as follows:

$$stab_{basic} = \frac{2 \times |E_i \cap E_j|}{|E_i| + |E_j|}$$

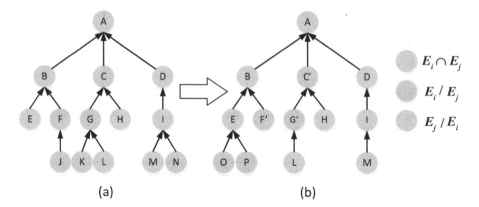

Fig. 4. Basic stability analysis

We take Fig. 4 as an example. Figure 4(a) and (b) refer to the two versions of an ontology before and after evolution. Three different colors respectively refer to $E_i \cap E_j$, E_i/E_j and E_j/E_i. And we can calculate: $|E_i| = 14$, $|E_j| = 13$, $|E_i \cap E_j| = 8$, $|E_i/E_j| = 6$, $|E_j/E_i| = 5$. So the basic stability: $stab_{basic} = \dfrac{2 \times 8}{14 + 13} \approx 0.593$.

The *Semantic stability* is:

$$stab_{semantic} = \frac{2 \times \left(|E_i \cap E_j| + \sum_{1}^{|E_{modify}|} sim \right)}{|E_i| + |E_j|}$$

Where $|E_{modify}|$ refers to the number of elements which are modified during evolution, and *sim* refers to the text similarity between modified elements and their corresponding elements before or after evolution.

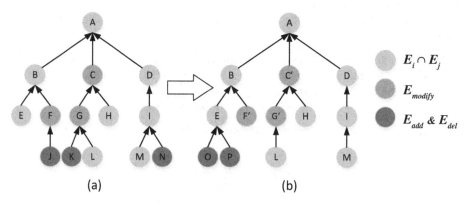

Fig. 5. Semantic stability analysis

We take Fig. 5 as an example. Three different colors respectively refer to $E_i \cap E_j$, E_{modify} and E_{add} & E_{del}. And we can calculate: $|E_i| = 14$, $|E_j| = 13$, $|E_i \cap E_j| = 8$, $|E_{modify}| = 3$, $|E_{del}| = 3$, $|E_{add}| = 2$. And the text similarities between the modified elements are: $Sim_{CC'} = 0.6$, $Sim_{FF'} = 0.7$, $Sim_{GG'} = 0.8$. So the semantic stability:

$$stab_{semantic} = \frac{2 \times (8 + (0.6 + 0.7 + 0.8))}{14 + 13} \approx 0.748.$$

We also propose the *Impact based on Gauss function*. According to *Gauss function* $f(x) = ae^{-(\frac{x-b}{\delta})^2}$, we proposed a kind of related measure between elements: $r_{ij} = r_0 \times e^{-(|E_i - E_j|)^2}$. r_{ij} refers to the related measure between E_i and E_j, r_0 refers to the constant and $|E_i - E_j|$ refers to the distance between two elements.

And on this basis, we discussed the impact brought by different change actions respectively.

- Ei is deleted, the impact: $imp_{del} = 1 + \sum_1^n r_{in}$, n refers to the number of element which is related to Ei before evolution (distance = 1).

- Ej is added, the impact: $imp_{add} = 1 + \sum_1^n r_{jn}$, n refers to the number of element which is related to Ej after evolution (distance = 1).

- Ei is modified to Ej, the impact: $imp_{modify} = s_{ij} \times (\sum_1^n r_{in} + \sum_1^m r_{jm})$, s_{ij} refers to the similarity between Ei and Ej (distance = 1).

So we can get the total impact based on *Gauss function*:

$$imp_{gauss} = \frac{\sum imp_{del}}{|E_i|} + \frac{\sum imp_{add}}{|E_j|} + \frac{\sum imp_{modify}}{|E_i| + |E_j|}$$

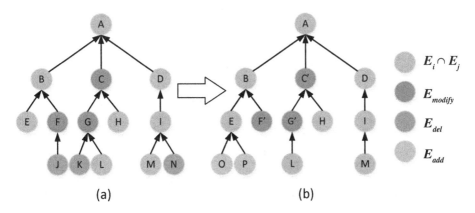

Fig. 6. Impact analysis based on *Gauss function*

We take Fig. 6 as an example. Four different colors respectively refer to $E_i \cap E_j$, E_{modify}, E_{del} and E_{add}. And we can calculate: $|E_i| = 14$, $|E_j| = 13$, $|E_i \cap E_j| = 8$, $|E_{modify}| = 3$, $|E_{del}| = 3$, $|E_{add}| = 2$. And the text similarities between the modified elements are: $Sim_{CC'} = 0.6$, $Sim_{FF'} = 0.7$, $Sim_{GG'} = 0.8$. We can calculate the impact of the deleted elements, added elements and modified elements:

$$\sum imp_{del} = (1 + e^{-1}) + (1 + e^{-1}) + (1 + e^{-1}) \approx 4.103,$$

$$\sum imp_{add} = (1 + e^{-1}) + (1 + e^{-1}) = 2.735,$$

$$\sum imp_{modify} = 0.6 \times 6 \times e^{-1} + 0.7 \times 3 \times e^{-1} + 0.8 \times 5 \times e^{-1} \approx 3.566.$$

So the total impact based on *Gauss function* is:

$$imp_{gauss} = \frac{\sum imp_{del}}{|E_i|} + \frac{\sum imp_{add}}{|E_j|} + \frac{\sum imp_{modify}}{|E_i| + |E_j|} \approx 0.636.$$

3   Results

OAEI 2007 benchmark is used here to validate our method. This data set contains more than 100 versions of an ontology, and is shown in Fig. 7.

All versions are organized as an evolution tree. Parameter d refers to evolution distance between two ontologies. The value range of d is in [1, 8] according to the evolution tree. In our experiments, for a given d, we randomly select 5 ontology pairs to obtain their evolution detection results and minimum cost evolution strategy, then we repeat the experiment 10 times and record the average evolution results. Therefore, we have 8 groups' results whose d is from 1 to 8 respectively.

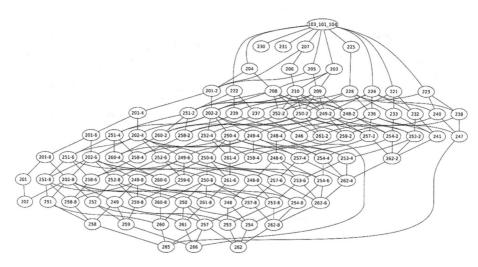

Fig. 7. OAEI 2007 benchmark

We obtain reference results manually with the help of the ontology edit tool Protégé, and the changes of concepts, properties, instances and all other elements between two ontologies are saved in a certain format. Precision, recall and F1-Measure are used to measure the experimental results.

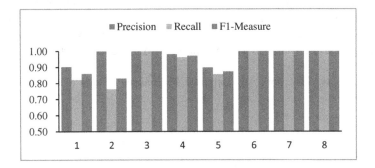

Fig. 8. Performance of concept evolution detection

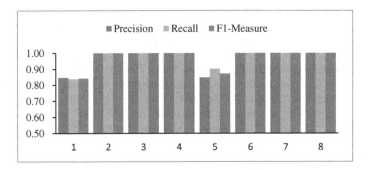

Fig. 9. Performance of property evolution detection

Figures 8 and 9 show the performances of evolution detection for concepts and properties, respectively. Wherein the horizontal axis represents the evolution distance d, and the y axis represents the precision, recall and F1-Measure. We can observe: (1) Our method can effectively detect ontology evolution; (2) With the evolution distance increasing to 6, our method becomes more effective.

The approach can be further subdivided into three sub-processes: ontology parse; constructing SDDs; TF-IDF calculation. According to the final results of the experiment, the similarity-based ontology evolution detection method has a good performance to capture the changes. Corresponding minimum cost evolution strategies can also be found by Algorithm 1, for the limitation of space, we will demonstrate them by our prototype system.

4 Conclusions

We propose a practical similarity-based ontology evolution detection method. Then the algorithm of finding the minimum cost strategy of ontology evolution is discussed. And we give three ideas to measure the evolution impact. Finally, we demonstrate the effectiveness of our method on OAEI benchmark data set. The proposed method would be useful for solving other relevant problems in ontology evolution, such as ontology matching evolution and reasoning update in ontology evolution. Besides, evolution impact analysis and evolution visualization are also our going work.

Acknowledgements. This work was supported by National Natural Science Foundation of China (61472077).

References

1. Stojanovic, L.: Methods and tools for ontology evolution. Ph.D. thesis, University of Karlsruhe (2004)
2. Noy, N.F., Klein, M.: Ontology evolution: not the same as schema evolution. Knowl. Inf. Syst. **6**(4), 328–440 (2004)
3. Groß, A., Hartung, M., Prüfer, K., et al.: Impact of ontology evolution on functional analyses. Bioinformatics **28**(20), 2671–2677 (2012)
4. Dong, G., Gao, Z., Qiu, X.: Automatic approach to ontology evolution based on change impact comparisons. Tsinghua Sci. Technol. **15**(6), 716–723 (2010)
5. Wang, P., Xu, B., Zhou, Y.: Extracting semantic subgraphs to capture the real meanings of ontology elements. J. Tsinghua Sci. Technol. **15**(6), 724–733 (2010)
6. Wang, P., Xu, B.: Lily: ontology alignment results for OAEI 2009. In: Proceedings of the Third International Workshop on Ontology Matching, pp. 167–175 (2008)
7. Wang, P., Zhou, Y., Xu, B.: Matching large ontologies based on reduction anchors. In: IJCAI, pp. 2343–2348 (2011)
8. Javed, M., Abgaz, Y.M., Pahl, C.: Composite ontology change operators and their customizable evolution strategies. In: The 2nd Joint Workshop on Knowledge Evolution and Ontology Dynamics EvoDyn 2012 - Collocated with the 11th International Semantic Web Conference (2012)
9. Noy, N.F., Chugh, A., Liu, W., Musen, M.A.: A framework for ontology evolution in collaborative environments. In: Cruz, I., Decker, S., Allemang, D., Preist, C., Schwabe, D., Mika, P., Uschold, M., Aroyo, L.M. (eds.) ISWC 2006. LNCS, vol. 4273, pp. 544–558. Springer, Heidelberg (2006)
10. Palma, R., Corcho, O., Haase, P., et al.: A holistic approach to collaborative ontology development based on change management. Web Semant. Sci. Serv. Agents World Wide Web **9**, 299–314 (2011)
11. Stojanovic, L., Maedche, A., Motik, B., Stojanovic, N.: User-driven ontology evolution management. In: Gómez-Pérez, A., Benjamins, V.R. (eds.) EKAW 2002. LNCS (LNAI), vol. 2473, pp. 285–300. Springer, Heidelberg (2002)
12. Groß, A., Hartung, M., Prüfer, K., et al.: Impact of ontology evolution on functional analyses. Bioinformatics **28**(20), 2671–2677 (2012)
13. Lee, S., Seo, W., Kang, D., et al.: A framework for supporting bottom-up ontology evolution. Expert Syst. Appl. **32**, 376–385 (2007)

14. Tury, M., Bieliková, M.: An approach to detection ontology changes. In: Workshop Proceedings of the Sixth International Conference on Web Engineering (2006)
15. Hartung, M., Groß, A., Rahm, E.: Conto–diff: Generation of complex evolution mappings for life science ontologies. J. Biomed. Inform. **46**(1), 15–32 (2013)

Ontology Matching Tuning Based on Particle Swarm Optimization: Preliminary Results

Pan Yang[1,2], Peng Wang[1,3(✉)], Li Ji[3], Xingyu Chen[3], Kai Huang[3], and Bin Yu[4]

[1] School of Computer Science and Engineering, Southeast University, Nanjing, China
{PanYoung,PWang}@seu.edu.cn
[2] School of Information Science and Engineering, Southeast University, Nanjing, China
[3] College of Software Engineering, Southeast University, Nanjing, China
[4] Communication Station of Unit 95028, P.L.A., Wuhan, China

Abstract. An ontology matching system can usually be run with different configurations to optimize the system's performance, namely precision, recall, or F-measure, depending on the given ontologies to be matched. Changing the configuration has potentially high impact on the obtained matching results. This paper applies particle swarm optimization to automatically tune these configuration parameters through proactively sampling the parameters space and find high-impact parameters and high-performance parameter settings. We show the effectiveness and efficiency of our approach through extensive evaluation on the OAEI 2009 tasks using Lily ontology matching system.

1 Introduction

Given a specific ontology matching task, an ontology matching system can usually be run with different configurations to optimize the system's performance. Changing the configuration has potentially high impact on the obtained results. Since ontology matching systems commonly have a large number of configuration parameters, regular users and even experts struggle to tune these parameters for good performance. This paper believe that automatic ontology matching techniques will increasingly supported by many complex ontology matching systems that use different strategies of combining multiple matching algorithms, which would consider one or more ontology features [1].

In this paper we propose an automatic ontology matching tuning method based on particle swarm optimization (PSO) [2]. Given a specific matching task, it will automatically take samples from the high-dimensional parameters space consisting of lots of configuration parameters and selects appropriate parameter settings to obtain the optimal matching results within an acceptable efficiency. Our approach is implemented in the Lily ontology matching system [3–5]. We first introduce the process of our automatic tuning work and its implementation in brief, and then demonstrate the advantages of this automatically configured

© Springer-Verlag Berlin Heidelberg 2014
D. Zhao et al. (Eds.): CSWS 2014, CCIS 480, pp. 146–155, 2014.
DOI: 10.1007/978-3-662-45495-4_13

system through evaluation against the datasets provided by the 2009 Ontology Alignment Evaluation Initiative (OAEI), compared with several other tuning techniques.

This paper is organized as follows: In Sect. 2, we will provide our framework for automatic ontology matching tuning for selecting best parameter configuration. Section 3 presents the experimental results and discusses the results. Section 4 is the related work. Conclusions are presented in Sect. 5.

2   Tuning to Automatically Select the Best Configuration

Our proposed matching process follows the steps shown in Fig. 1. First, one pair of ontologies to be matched is loaded into the ontology matching module and the automatic tuning module gets initialized using Latin Hypercube Sampling (LHS) [6]. The matching operation is performed upon one sample of configuration is taken by the PSO algorithm from the high-dimensional parameters space and be passed into the ontology matching module through the data interaction module. Then the automatic tuning module gets the corresponding matching results as feedback to direct its next sampling after the operation. Finally, within limited number of iterations described as above, the optimal configuration is recommended to users and the corresponding alignment is obtained. We also make use of obtained empirical results to analyze each parameter's importance, in order to improve the original tuning work.

In our implementation, we first adopt LHS to initialize the original PSO algorithm. Since size of the particle swarm is predefined as 10, LHS is supposed to take a set of 10 samples as each particle's initial position respectively, from the parameters space at one time. For the reason that LHS itself does not rule out bad spreads (for example, all samples spread along the diagonal), we address this problem by generating 1000 sets of LHS samples and finally choosing the one

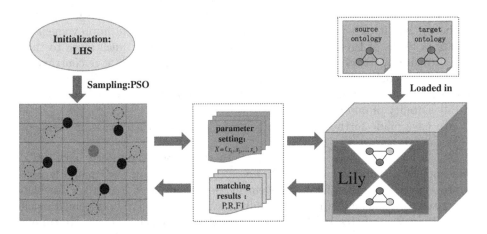

Fig. 1. Automatic configuration parameter tuning process

that maximize the minimum distance between any path of samples. After initialization, it begins to iteratively search for the optimal configurations using the PSO algorithm. After finishing 10 iterations, the tuning work gets stop and gives the optimal parameter configuration to users with the corresponding alignment.

3   Evaluation and Results

In this section we describe the experimental results obtained using the approach introduced in Sect. 2. Our experiments have been run using 127 matching tasks from three tracks (namely benchmark, conference and anatomy) with reference alignments provided in OAEI 2009. Tuning tasks in our evaluation consider up to 22 configuration parameters belonging to two matcher for generic and large ontology matching tasks respectively. And we study the impact of our approach in terms of Precision, Recall and F-measure respectively.

Our evaluation compares the proposed approach, labeled *LHS-PSO*, against:

(1) **Default** parameter settings that come with the system Lily.

(2) **Random-PSO** tuning gets initialized by Random Sampling, instead of LHS. Then continuing tuning work using PSO algorithm within 10 particles in size and 10 iterations in implementation, like LHS-PSO.

(3) **Approximation to the optimal setting:** since we do not know the optimal performance in any tuning scenario, we run large quantities of experiments for each tuning task. We have done 100 experiments in random per tuning task, which equals to the overall experiments of LHS-PSO or Random-PSO per tuning task. The best performance found is used as an approximation of the optimal. This technique is labeled *Brute Force*.

As shown in Fig. 2, in most cases *Random-PSO* outperformed other techniques and made large improvement, more than 10 % in average, in terms of F-measure compared to the default parameter settings'. *Brute Force* also obtained good results, which is even slightly better than that of *Random-PSO* in terms of F-measure in track of anatomy. *LHS-PSO* only outperformed the tuning work using the default parameter settings, instead of being the optimal one as expected in theory, but it also made large improvement. In detail, the improvement of results in terms of Precision, instead of Recall, contributes much more to that of F-measure in tracks of both anatomy and conference, while it is contrary in the case of benchmark, no matter which tuning technique used. These results have shown the feasibility and validity of our proposed method and the feature of Lily's ability of ontology matching.

We have made further analysis about the results of each track, in terms of F-measure.

As shown in Fig. 3, almost all tuning methods performed perfectly with 100 % achievement in sub-tracks of 101–104 & 221–247. *Random-PSO* made largest improvement, more than 20 % achievement, in sub-tracks of 201–210 & 248–266 compared to the default. In the same sub-tracks, *Random-PSO* performed much better than *Brute Force* & *LHS-PSO*, nearly 10 % more achievement. All tuning methods performed not well in the sub-track of 301–304, as well as the sub-track of 248–266, which indicates the hardness of tuning tasks in these sub-tracks.

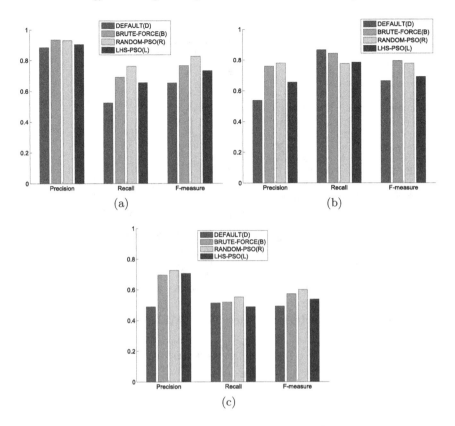

Fig. 2. Results on OAEI 2009 benchmark, anatomy & conference tasks

As for the track of anatomy, we have adopted two different optimal results in previous OAEI as references, namely *amaker_1* & *SOBOM_1*, since the standard references are not open to public. As shown in Fig. 4, although references are not the same, the results on this track are almost the same to all tuning methods. Concerning the track of conference, we can find easily that *Random-PSO* made largest improvement in task of cmt-edas, more than 20 % achievement, compared to the default, which is the same to *Brute-Force*. In task of confOf-sigkdd, *Random-PSO* also made great improvement, nearly 20 % more achievement than all other methods. However, all tuning methods performed not well all over this track, which is obvious in Fig. 5.

In order to improve the original tuning work, we have made further experiments to analyze the importance of each configuration parameter.

According to Fig. 4, it is obvious that there is only one high-impact parameter, namely *ntValue*, in tasks of anatomy. In detail, we can find that other two parameters, namely *ptValue* and *nbScale*, also have considerable impact on the results, although not such high like that of parameter *ntValue*.

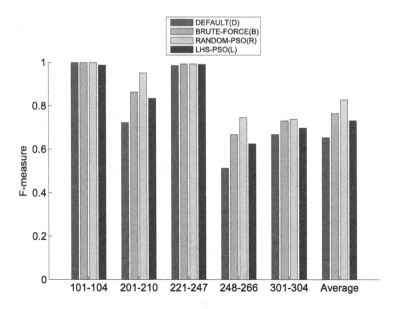

Fig. 3. Detail analysis of the results on OAEI 2009 benchmark

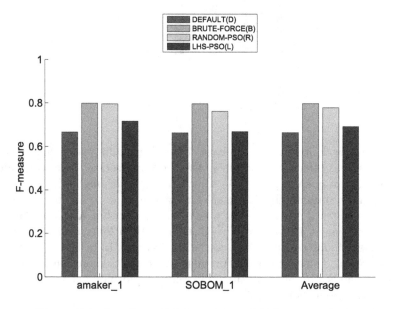

Fig. 4. Detail analysis of the results on OAEI 2009 anatomy

According to Fig. 5, although the number and type of main high-impact para-meters (namely *cnptSim Threshold, propSim Threshold, simProgType, edThresh-old*) are still the same in these tasks, their sensitive curves have changed greatly for each different task.

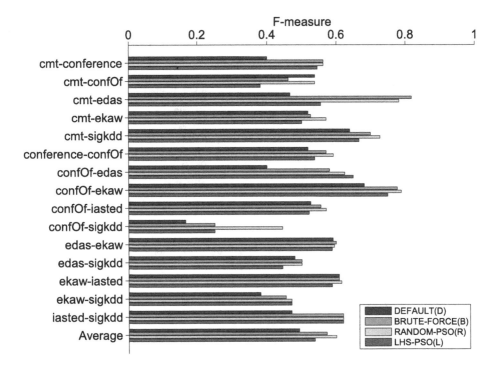

Fig. 5. Detail analysis of the results on OAEI 2009 conference

Through this analysis, we can clearly separate the important parameters from the unimportant ones, which can be set to their default values with little negative on performance, and need not to be considered for tuning. Furthermore, since only a small part of parameters are high-impact, compared to 22 configuration parameters overall, *LHS-PSO* did not reach its greatest potential to achieve the best performance as expected in theory, in limited number of iterations as shown in Fig. 2.

4   Related Work

In database field, tuning parameter is important for performance of database systems and schema matching systems. Lee et al. propose eTuner, which is an automatic tuning method for schema matching in database field [7]. eTuner exploits the synthetic scenario to generate a set of synthetic schema and corresponding matchings, then finds the good tuning configuration based on the synthetic schema and matchings. A hierarchy tuning strategy is also used to reduce the searching space during tuning. However, the synthetic scenario would be different to the real-world applications. iTuned system proposes an adaptive sampling technique to realize the automatic tuning for the amount of parameters in database systems [8,9]. This method tries to find the parameters with higher

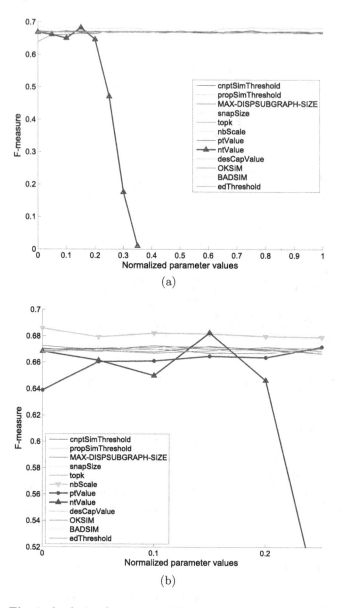

Fig. 6. Analysis of parameters' importance on anatomy task

impact, then recommend these parameters to database systems. However, the tuning performance of this method is dependent on the initial sampling data. Peukert et al. propose a self-configuration schema matching system, which uses various features of shema and middle matching results, then some rules are used to realize the tuning for specific matching tasks [10] (Figs. 6 and 7).

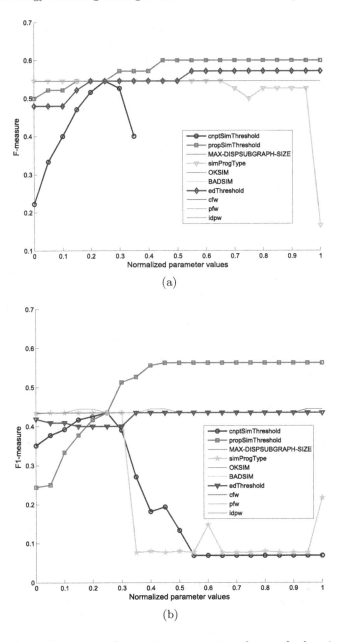

Fig. 7. Analysis of parameters' importance on cmt-conference & ekaw-iasted tasks

Shvaiko et al. pointed out that ontology matching tuning is one of the challenges in ontology matching [11]. Till now, some works have been done on this problem. Some works focus on the matcher selection problem in matching tuning. Mochol et al. propose a rule-based matcher selection algorithm, which recommends the suitable matchers according to the performances of matchers on different matching tasks [12]. Eckert et al. propose a classification method to learn correct combinations of matchers [13]. The classifier uses the features such as lexical and structural features of ontology concepts, and simple ontology features. ECOMatch is one automatic parameter tuning system for ontology matching [14]. It uses the reference alignments provided by users as the input, then meta-heuristics and machine learning methods are used to search the huge space of parameters. Cruz et al. proposed a learning-based configuration selection method [1], which requires training dataset including matching tasks and reference alignments. Bock et al. apply PSO to ontology matching, and propose new ontology matching algorithm MapPSO [15]. This method treats the ontology matching problem as an optimization problem.

5   Conclusions

We proposed an improved PSO algorithm to automatically search for the best configuration from the space of possible configurations. We also introduced well-designed experiments to analyze configuration parameters' importance in real tasks. Although our approach is evaluated using the Lily ontology matching system only, it can be implemented easily for any other ontology matching systems. Our work can be also extended by adopting other tuning techniques or related search algorithms and modeling the parameters space using obtained empirical results to improve the process of sampling [8]. Another interesting research topic consists of analyzing the relations among parameters.

Acknowledgments. This work was supported by the National Natural Science Foundation of China (61472077).

References

1. Cruz, I.F., Fabiani, A., Caimi, F., Stroe, C., Palmonari, M.: Automatic configuration selection using ontology matching task profiling. In: Simperl, E., Cimiano, P., Polleres, A., Corcho, O., Presutti, V. (eds.) ESWC 2012. LNCS, vol. 7295, pp. 179–194. Springer, Heidelberg (2012)
2. Kennedy, J., Eberhart, R., et al.: Particle swarm optimization. In: Proceedings of IEEE International Conference on Neural Networks, pp. 1942–1948 (1995)
3. Wang, P., Xu, B.: Lily: ontology alignment results for OAEI 2008. In: Proceedings of the 3rd International Workshop on Ontology Matching, pp. 167–175 (2008)
4. Wang, P.: Lily results on SEALS platform for OAEI. In: Proceedings of the 6th International Workshop on Ontology Matching, pp. 156–162 (2011)
5. Wang, P., Zhou, Y., Xu, B.: Matching large ontologies based on reduction anchors. In: Proceedings of the 22nd International Joint Conference on Artificial Intelligence (IJCAI'11), pp. 2343–2348 (2011)
6. Hicks, C.R., Turner, K.V.: Fundamental concepts in the design of experiments (1999)

7. Lee, Y., Sayyadian, M., Doan, A., et al.: eTuner: tuning schema matching software using synthetic scenarios. VLDB J. - Int. J. Very Large Data Bases **16**(1), 97–122 (2007)
8. Duan, S., Thummala, V., Babu, S.: Tuning database configuration parameters with iTuned. Proc. VLDB Endowment **2**(1), 1246–1257 (2009)
9. Thummala, V., Babu, S.: iTuned: a tool for configuring and visualizing database parameters. In: Proceedings of the 2010 ACM SIGMOD International Conference on Management of Data, pp. 1231–1234 (2010)
10. Peukert, E., Eberius, J., Rahm, E.: A self-configuring schema matching system. In: 2012 IEEE 28th International Conference on Data Engineering (ICDE), pp. 306–317 (2012)
11. Shvaiko, P., Euzenat, J.: Ten challenges for ontology matching. In: Meersman, R., Tari, Z. (eds.) OTM 2008, Part II. LNCS, vol. 5332, pp. 1164–1182. Springer, Heidelberg (2008)
12. Mochol, M., Jentzsch, A.: Towards a rule-based matcher selection. In: Gangemi, A., Euzenat, J. (eds.) EKAW 2008. LNCS (LNAI), vol. 5268, pp. 109–119. Springer, Heidelberg (2008)
13. Eckert, K., Meilicke, C., Stuckenschmidt, H.: Improving ontology matching using meta-level learning. In: Aroyo, L., Traverso, P., Ciravegna, F., Cimiano, P., Heath, T., Hyvönen, E., Mizoguchi, R., Oren, E., Sabou, M., Simperl, E. (eds.) ESWC 2009. LNCS, vol. 5554, pp. 158–172. Springer, Heidelberg (2009)
14. Zhou, Y.: Extensions of an empirical automated tuning framework. Master Thesis. University of Maryland, College Park (2013)
15. Bock, J., Hettenhausen, J.: Discrete particle swarm optimisation for ontology alignment. Inf. Sci. **192**, 152–173 (2012)

AxiomVis: An Axiom-Oriented Ontology Visualization Tool

Jian Zhou[1,2], Xin Wang[1,2(✉)], and Zhiyong Feng[1,2]

[1] School of Computer Science and Technology, Tianjin University, Tianjin, China
[2] Tianjin Key Laboratory of Cognitive Computing and Application, Tianjin, China
{jianzhou,wangx,zyfeng}@tju.edu.cn

Abstract. With the continuous growth of ontologies in size, the amount of ontology axioms is steadily increasing. To help users understand ontologies with a large number of axioms, we devise a novel ontology visualization tool, called *AxiomVis*, which defines a mapping from DL axioms to the *directed acyclic graph* (DAG). We give a recursive procedure to generate the target DAG and then propose a new *axiom-aware* layout algorithm to display the DAG elements using the characteristics of the expression tree. The distinguishing feature of our tool is that it visualizes ontologies from the axiom viewpoint. This paper demonstrates comparison of different layouts, which exhibits the advantages of our layout on ontology visualization.

Keywords: Ontology · Axioms · Visualization · Layout

1 Introduction

With the rapid development of the Semantic Web, more and more ontologies have been released in the Web Ontology Language (OWL) as standardized by W3C [1]. As a cornerstone of OWL, description logics are a family of logic-based knowledge representation formalisms, which support the logic description of concepts, roles, and their relationships, using a variety of operators, to form more complex descriptions. A description logic knowledge base typically consists of two components, i.e., a TBox and an ABox. In a TBox, the concept definition of a new concept is in terms of other defined concepts. For example, the fact that orphan's parents are not alive is expressed by the concept inclusion

$$\texttt{Orphan} \sqsubseteq \texttt{Human} \sqcap \forall\texttt{hasParent}.\neg\texttt{Alive} \tag{1}$$

in which case we say that the concept orphan is subsumed by the composite concept

$$\texttt{Human} \sqcap \forall\texttt{hasParent}.\neg\texttt{Alive} \tag{2}$$

This paper focuses on TBox axioms since they are more complex than ABox axioms and our techniques can be easily adapted to handle ABox axioms (Note that in the following we use the term *axiom* to refer to a TBox axiom).

© Springer-Verlag Berlin Heidelberg 2014
D. Zhao et al. (Eds.): CSWS 2014, CCIS 480, pp. 156–166, 2014.
DOI: 10.1007/978-3-662-45495-4_14

It is known that axioms are the first-class citizens in ontologies. However, The majority of the existing techniques for ontology visualization [2–12] are not aware of axioms, which cannot represent relationships among concepts, roles, and operators from the axiom perspective. For example, several visualization techniques include OWL-VisMod [14] and Prefuse [15], which use treemapping technology that is a method for displaying hierarchical data by using nested rectangles to visualize certain aspects of ontologies. While these techniques consider the class hierarchy of ontologies, they do not take into account axioms. The approaches that are closely related to AxiomVis include GrOWL [13] and SOVA[1], which use graph-based visualization techniques to represent ontologies. GrOWL and SOVA define more elaborated notations in order to be consistent to ontology semantics according to the custom criteria. However, considering the fact that description logics are decidable subsets of first-order logic, we can utilize the form of the expression tree to represent ontologies. We propose an axiom-aware ontology visualization tool, called AxiomVis, which (1) constructs a DAG according to a set of axioms, (2) draws a DAG according to our axiom-aware layout, and (3) shows a set of demonstration use cases.

2   Algorithms

In this section, we describe the core algorithms used in AxiomVis, which can construct a DAG from axioms and then draw this graph according to our axiom-aware layout.

We present the DAG construction function $dag : \mathcal{P}(TB) \rightarrow DAG$, where $\mathcal{P}(TB)$ is the power set of the set of axioms TB and DAG is the set of DAGs. Let N_C and N_R be infinite sets of concept and role names, respectively. We define the set of concept descriptions on N_C and N_R as $Desc$ that is the smallest set such that (1) for each $A \in N_C$, $A \in Desc$, (2) if $C \in Desc$ and $D \in Desc$, then $\neg C \in Desc$, $C \sqcap D \in Desc$, and $C \sqcup D \in Desc$, and (3) if $r \in N_R$ and $C \in Desc$, then $\exists r.C \in Desc$ and $\forall r.C \in Desc$. A TBox axiom is of the form $A \equiv C$ or $C \sqsubseteq D$, where $A \in N_C$, $C \in Desc$, and $D \in Desc$. We use $O = \{\neg, \sqcap, \sqcup, \exists, \forall, \equiv, \sqsubseteq\}$ to denote the set of operators. TB can be modelled as a DAG $G = (V, E)$, where the node set $V = \{v \mid v \in N_C \cup N_R \cup O\}$, the edge set $E \subseteq V \times V$. Assume the function $root : Desc \cup N_R \rightarrow V$, where this function returns the root node of the expression tree that corresponds to $d \in Desc$. If $d \in N_C \cup N_R$, then $root(d) = d$. For a given set of axioms TB, the function $dag(TB)$ constructs a DAG G in accordance with the following steps: for each $t \in TB$, (1) if $\neg C \in Desc$ and $\neg C$ occurs in T, there exists an edge $e = (\neg, root(C)) \in E$, (2) if $C \in Desc$, $D \in Desc$, $m \in \{\sqcap, \sqcup \sqsubseteq, \equiv\}$, and CmD occurs in T, there exists two edges $e_1 = (m, root(C)) \in E$ and $e_2 = (m, root(D)) \in E$, and (3) if $r \in N_R$, $C \in Desc$, $m \in \{\exists, \forall\}$, and $mr.C$ occurs in t, there exists two edges $e_1 = (m, r) \in E$ and $e_2 = (m, root(C)) \in E$. When the number of axioms is only one, the DAG G degenerates into a expression tree. For example, the axiom (1) can be converted

---

[1] http://protegewiki.stanford.edu/wiki/SOVA

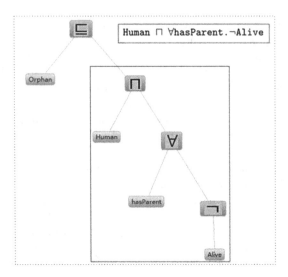

Fig. 1. The subtree representation of the composite concept

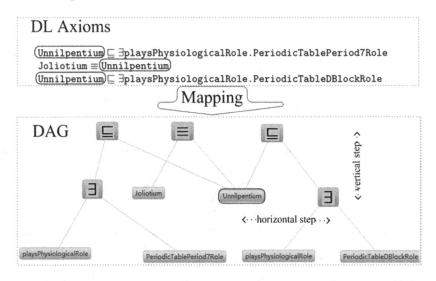

Fig. 2. A mapping from DL axioms to the DAG and axiom-aware layout

to the expression tree, and the composite concept (2) can be converted to the subtree as shown in Fig. 1, whose root node is the operator $\sqsubseteq$.

We design the aware-axiom layout algorithm in order to draw the DAG G. Each axiom is an expression tree in the DAG. By calculating the depth of the expression tree and the number of nodes in each layer, we can easily get the horizontal step for the tree and the vertical step for each layer as shown in Fig. 2. The vertical step and the horizontal step determine the vertical coordinates of

Algorithm 1. Axiom-aware layout

Input: the set of axioms TB
Output: the DAG G, coordinates of each node in the DAG G
1: **for** each $t \in TB$ **do**
2:     $G \leftarrow dag(TB)$
3:     $wd \leftarrow w/|TB|$ $\triangleright$ wd is the horizontal width of a expression tree in the canvas
4:     $k \leftarrow 1$, $x \leftarrow 0$, $y \leftarrow 0$, $prex \leftarrow 0$ $\triangleright$ Coordinates of node (x, y)
5:     $et \leftarrow G.getTree(t)$
6:     $vs \leftarrow h/(et.layer + 1)$ $\triangleright$ vs is the vertical step of the adjacent layer
7:     **while** $k \leqslant et.layer$ **do**
8:         $Q \leftarrow et.getQueue(k)$
9:         $hs \leftarrow wd/(Q.size() + 1)$ $\triangleright$ hs is the horizontal step of the same layer
10:         **while** $!Q.isEmpty()$ **do**
11:             $node \leftarrow Q.dequeue()$
12:             $x \leftarrow x + hs$ $\triangleright$ Update the horizontal coordinates of nodes
13:             $node.setLoc(x, y)$
14:         **end while**
15:         $y \leftarrow y + vs$ $\triangleright$ Update the vertical coordinates of nodes
16:         $k \leftarrow k + 1$ $\triangleright$ Move down to the next layer
17:     **end while**
18:     $x \leftarrow prex + wd$ $\triangleright$ Update the horizontal coordinate of the root node
19:     $y \leftarrow 0$ $\triangleright$ Update the vertical coordinate of the root node
20:     $prex \leftarrow x$
21: **end for**

nodes between the adjacent layers and the horizontal coordinates of nodes in the same layer, respectively. For the sake of managing the position of each node in the tree, multiple queues are created. The number of queues is the depth of the tree. Each queue allows to enqueue the nodes with the same layer in the tree. When drawing the node in the DAG, each queue keeps the horizontal step to dequeue the nodes with the same layer, and then the queues maintain the vertical step between the adjacent layers. Let h be the height of the canvas, w be the width of the canvas, et be a expression tree, $et.layer$ be the layer of et, and Q is a queue that is composed by the nodes with the same layer in et. Algorithm 1 shows the procedure for the axiom-aware layout, in which the function $G.getTree$, $et.getQueue$, and $node.setLoc$ are used to get the expression tree et in G according to the given axiom t, get a queue Q for one layer k in et, and set the location of $node$ in the canvas, respectively.

We establish a mapping from DL axioms to the DAG. For example, we visualize all DL axioms that include the concept Unnilpentium in OpenGALEN. It is obvious to observe that the node Unnilpentium in the DAG represents the concept in DL axioms and a tree in the DAG denotes an axiom as shown in Fig. 2.

3   Demonstration

This section presents the demonstration environment and scenarios of AxiomVis, which is published as an open source software[2].

3.1   Demonstration Environment

AxiomVis is written in Java, and the version of JDK is 1.7. The Eclipse Rich Client Platform (RCP) as the basis of AxiomVis. The visualization part of AxiomVis is developed using Zest[3], a set of visualization components built for Eclipse. The main feature of Zest is that it makes graph programming easy. Our computer has Intel 3.30 GHz CPU and 4 GB memory. The operating system is Windows 7. Our method is general enough to be applied to all OWL ontologies. Due to space limitations, we only use OpenGALEN[4] (209,248 axioms, 84 fragments of OWL) as an example to demonstrate the functionalities of our AxiomVis.

3.2   Demonstration Scenarios

In this section, we present the demonstration scenarios by the following use cases.

Demonstration use case 1: Querying the number of the concepts or roles.

When choosing 8 different ontology fragments in OpenGALEN, we can get the total number of the concepts or roles in the selected fragment. Table 1 shows these query results.

Table 1. Demonstration results of querying the number of the concepts or roles

Serial number	Ontology fragment	#concepts	#roles
t1	FoundationModel_ClinicalSituationModel	4090	1838
t2	MedicalExtensions_HumanAnatomy	11682	1853
t3	MedicalExtensions_Genetics	11631	1229
t4	DissectionsDisease	34342	5514
t5	DS_4076_Musculoskeletal	37148	6436
t6	DS_85_Diagnostic	27764	2209
t7	DS_3227_Cardiothoracic	34054	5048
t8	DD_8048_Components	37667	5054

Demonstration use case 2: Querying a concept or role.

---

[2] AxiomVis. http://xinwang.tju.edu.cn/drupal/?q=research/demo/axiomvis.
[3] Zest. http://www.eclipse.org/gef/zest/.
[4] OpenGALEN. http://www.opengalen.org/.

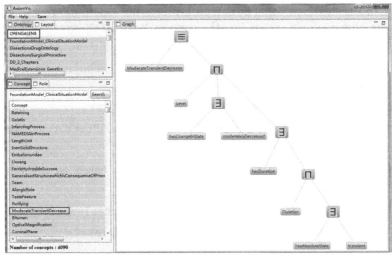

(a) Ontology visualization according to the selected concept

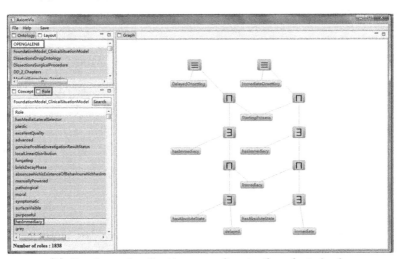

(b) Ontology visualization according to the selected role

Fig. 3. Screenshots of demonstration use cases 3

If you want to query a concept or role in the selected fragment of OWL, the query results will be listed after clicking on the search button in the fragment FoundationModel_ClinicalSituationModel as shown in Fig. 6(a). We can also obtain the total number of the concept or role in this fragment.

Demonstration use case 3: Ontology visualization according to the selected concept or role.

We can get visualization information, for instance, if we choose the concept ModerateTransientDecrease or the role hasImmediacy in OpenGALEN as shown in Fig. 3(a) and (b). The default layout is our axiom-aware layout. It is easy to get the axiom (3) related to the concept `ModerateTransientDecrease` from Fig. 3(a).

> `ModerateTransientDecrease` ≡
>> `Level` ⊓
>> `∃hasChangeInState.moderatelyDecreased` ⊓
>> `∃hasDuration.(Duration` ⊓ `∃hasAbsoluteState.transient)`
>
> (3)

Similarly, we can obtain the following two axioms based on the role `has Immediacy` from Fig. 3(b).

> `DelayedOnsetting` ≡ `StartingProcess` ⊓ `∃hasImmediacy.`
>> `(Immediacy` ⊓ `∃hasAbsoluteState.delayed)`   (4)

> `ImmediateOnsetting` ≡ `StartingProcess` ⊓ `∃hasImmediacy.`
>> `(Immediacy` ⊓ `∃hasAbsoluteState.immediate)`   (5)

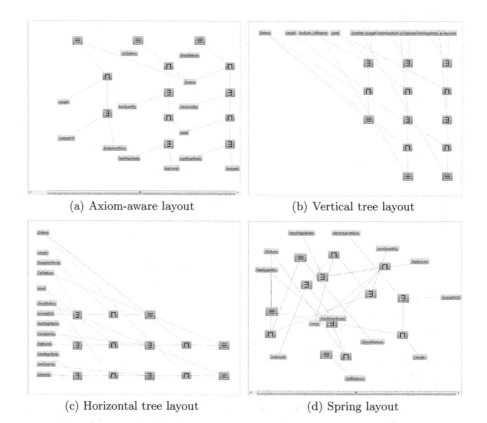

(a) Axiom-aware layout   (b) Vertical tree layout

(c) Horizontal tree layout   (d) Spring layout

Fig. 4. Screenshots of demonstration use case 5

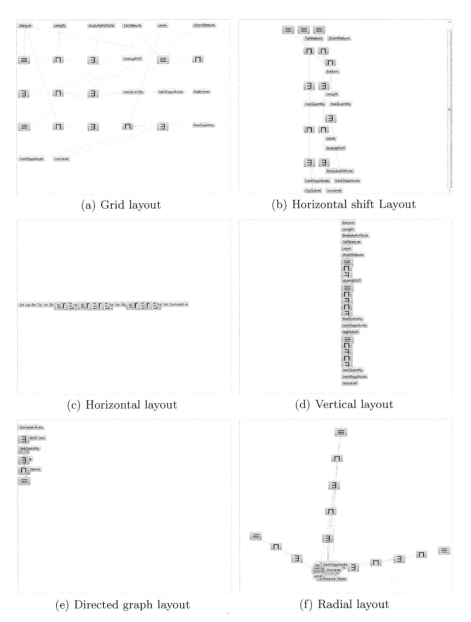

(a) Grid layout

(b) Horizontal shift Layout

(c) Horizontal layout

(d) Vertical layout

(e) Directed graph layout

(f) Radial layout

Fig. 5. Screenshots of demonstration use case 5

If necessary, we also can switch to a different Layout, drag the nodes and edges in the canvas, and save this canvas.

Demonstration use case 4: Graph layout options.

If you want to choose a different layout for the current visual graph. AxiomVis provides a number of layout options including our axiom-aware layout and 9 layouts from Zest as shown in Fig. 6(b).

Demonstration use case 5: Comparison of layout algorithms.

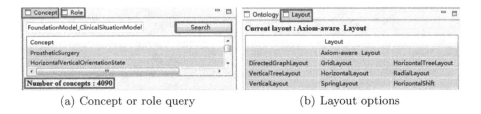

(a) Concept or role query   (b) Layout options

Fig. 6. Screenshots of demonstration use cases 2, 4

We choose our axiom-aware layout algorithm. As a comparison, We also select 9 layouts from Zest that include VerticalTreeLayout, HorizontalTreeLayout, SpringLayout, GridLayout, HorizontalShiftLayout, HorizontalLayout, VerticalLayout, DirectedGraphLayout, and RadialLayout for ontology visualization. Figures 4(a)–(d) and 5(a)–(f) show the results of clicking the concept `Stature` in OpenGALEN. HorizontalTreeLayout and VerticalTreeLayout cannot clearly distinguish each axiom in the canvas, although they can keep a tree structure from the horizontal or vertical direction. SpringLayout, GridLayout, and HorizontalShiftLayout cannot make full use of the characteristics of the expression tree for each axiom even though they can properly arrange the position of each node in the DAG. HorizontalLayout, VerticalLayout, and DirectedGraphLayout pay too much attention to the overall layout of direction, which lead to the overcrowding of visual elements. RadialLayout selects some elements as the root nodes, and then arranges the other nodes to form a series of circles, which also ignores the characteristics of the expression tree. In summary, it can be observed that our layout has more advantages on dealing with ontology visualization than the state-of-the-art layouts.

4   Conclusion

This paper presents a new axiom-oriented visualization approach, which can facilitate the analysis and understanding of ontologies from the axiom perspective. In the future, many aspects of the approach can be improved including the design of user-friendly interface and further optimization of axiom-aware layout algorithm. We hope to integrate more features into AxiomVis. In particular, we would like to improve the performance of AxiomVis and design a hierarchical structure to visualize large-scale ontologies. We also need to address the representation of individuals and extend our algorithms to handle ABox axioms.

Acknowledgments. This work is supported by the National Natural Science Foundation of China (61100049), the Graduate English Course Construction Project of Tianjin University (S216E003), and the National High-tech R&D Program of China (863 Program) (2013AA013204).

References

1. McGuinness, D.L., Van Harmelen, F.: OWL web ontology language overview. W3C recommendation (2004)
2. Alani, H.: TGVizTab: an ontology visualisation extension for Protégé. In: Knowledge Capture Workshop on Visualization Information in Knowledge Engineering (2003)
3. Knublauch, H., Fergerson, R.W., Noy, N.F., Musen, M.A.: The Protégé OWL plugin: an open development environment for semantic web applications. In: McIlraith, S.A., Plexousakis, D., van Harmelen, F. (eds.) ISWC 2004. LNCS, vol. 3298, pp. 229–243. Springer, Heidelberg (2004)
4. Hussain, A., Latif, K., Rextin, A., Hayat, A., Alam, M.: Scalable visualization of semantic nets using power-law graphs. Appl. Math. Inf. Sci. 8(1), 355–367 (2014)
5. Negru, S., Haag, F., Lohmann, S.: Towards a unified visual notation for OWL ontologies: insights from a comparative user study. In: 9th International Conference on Semantic Systems, pp. 73–80 (2013)
6. Motta, E., Mulholland, P., Peroni, S., d'Aquin, M., Gomez-Perez, J.M., Mendez, V., Zablith, F.: A novel approach to visualizing and navigating ontologies. In: Aroyo, L., Welty, C., Alani, H., Taylor, J., Bernstein, A., Kagal, L., Noy, N., Blomqvist, E. (eds.) ISWC 2011, Part I. LNCS, vol. 7031, pp. 470–486. Springer, Heidelberg (2011)
7. Heim, P., Lohmann, S., Stegemann, T.: Interactive relationship discovery via the semantic web. In: Aroyo, L., Antoniou, G., Hyvönen, E., ten Teije, A., Stuckenschmidt, H., Cabral, L., Tudorache, T. (eds.) ESWC 2010, Part I. LNCS, vol. 6088, pp. 303–317. Springer, Heidelberg (2010)
8. Lanzenberger, M., Sampson, J.: Alviz - a tool for visual ontology alignment. In: 10th International Conference on Information Visualization, pp. 430–440 (2006)
9. Storey, M.-A., Noy, N.F., Musen, M., Best, C., Fergerson, R., Ernst, N.: Jambalaya: interactive visualization to enhance ontology authoring and knowledge acquisition in protégé. In: Workshop on Interactive Tools for Knowledge Capture (2001)
10. Katifori, A., Torou, E., Halatsis, C., Lepouras, G., Vassilaki, C.: A comparative study of four ontology visualization techniques in protégé: experiment setup and preliminary results. In: Information Visualization, pp. 417–423. IEEE (2006)
11. Parsia, B., Wang, T., Golbeck, J.: Visualizing web ontologies with cropcircles. In: 4th International Semantic Web Conference, pp. 6–10 (2005)
12. Wang, T.D., Parsia, B.: CropCircles: topology sensitive visualization of OWL class hierarchies. In: Cruz, I., Decker, S., Allemang, D., Preist, C., Schwabe, D., Mika, P., Uschold, M., Aroyo, L.M. (eds.) ISWC 2006. LNCS, vol. 4273, pp. 695–708. Springer, Heidelberg (2006)
13. Krivov, S., Villa, F., Williams, R., Wu, X.: On visualization of OWL ontologies. In: Baker, C.J.O., Cheung, K.-H. (eds.) Semantic Web, pp. 205–221. Springer, New York (2007)

14. García-Peñalvo, F.J., Colomo-Palacios, R., García, J., Therón, R.: Towards an ontology modeling tool. A validation in software engineering scenarios. Expert Syst. Appl. **39**(13), 11468–11478 (2012)
15. Heer, J., Card, S., Landay, J.: Prefuse: a toolkit for interactive information visualization. In: Proceedings of the SIGCHI Conference on Human factors in Computing Systems, pp. 421–430 (2005)

Exploiting Semantic Web Datasets: A Graph Pattern Based Approach

Honghan Wu[1,3]([⊠]), Boris Villazon-Terrazas[2], Jeff Z. Pan[1],
and Jose Manuel Gomez-Perez[2]

[1] Department of Computing Science, University of Aberdeen, Aberdeen, UK
honghan.wu@abdn.ac.uk
[2] ISOCO, Intelligent Software Components S.A., Madrid, Spain
[3] Nanjing University of Information Science and Technology, Nanjing, China

Abstract. In the last years, we have witnessed vast increase of Linked Data datasets not only in the volume, but also in number of various domains and across different sectors. However, due to the nature and techniques used within Linked Data, it is non-trivial work for normal users to quickly understand what is within the datasets, and even for tech-users to efficiently exploit the datasets. In this paper, we propose a graph pattern based framework for realising a customisable data exploitation. Atomic graph patterns are identified as building blocks to construct facilities in various exploitation scenarios. In particular, we demonstrate how such graph patterns can facilitate quick understandings about RDF datasets as well as how they can be utilised to help data exploitation tasks like concept level browsing, query generation and data enrichment.

1 Introduction

So far, Linked Data principles and practices are being adopted by an increasing number of data providers, getting as result a global data space on the Web containing hundreds of LOD datasets [2]. However the technical prerequisites of using Semantic Web dataset prevent efficient exploitations on these datasets. To tackle this problem, in this paper we present a summarisation based approach which can not only provide a quick understanding of the dataset in question, but also is able to guide users in exploiting it in various ways.

In this demo, we will introduce our graph pattern based exploitation system[1] and demonstrate three exploitation tasks of *(Quick Understanding)* big picture presenting and summary browsing, *(Guided Exploitation)* two query generation methods, and *(Dataset Enrichment)* atomic pattern based dataset linkage.

The rest of paper is organized as follows: in Sect. 2, we briefly discuss the related work. Then, in Sect. 3 we introduce the details of the summarisation definition and generation. In Sect. 4, we demonstrate several typical data exploitation scenarios based on the information and properties of our summarisation. Conclusions and future work are briefly given in the final section.

---

[1] http://homepages.abdn.ac.uk/honghan.wu/pages/kd.wp3/

© Springer-Verlag Berlin Heidelberg 2014
D. Zhao et al. (Eds.): CSWS 2014, CCIS 480, pp. 167–173, 2014.
DOI: 10.1007/978-3-662-45495-4_15

2   Related Work

In this section we describe the related work regarding the consumption and exploitation of Linked Data datasets. We split the related works in two groups (1) works that deals with the extraction of metadata from the datasets; and (2) works related to dataset summarization.

In the following we are going to briefly the related works that extract metadata from available datasets

- *LODStats* [3] provides detailed information of LOD datasets such as structure, coverage, and coherence, for helping users to reuse, link, revise or query dataset published on the Web. It is a statement-stream-based approach for collecting statistics about datasets described in RDF. *LODStats* have smaller memory foodprint, and better performance and scalability.
- *make-void*[2] is another tool that computes statistics for RDF datasets and generates RDF output using the VoID vocabulary. It is based on Jena and features advanced criteria, such as the number of links between URI namespaces, or the number of distinct subjects.
- Holst [5] provides an automated structural analysis of RDF data based on SPARQL queries. He identifies, first, a set of 18 measures for structural analysis. Next, he implements the measures in his RDFSynopsis tool, following two approaches (1) Specific Query Approach (SQA) and Triple Stream Approach (TSA). Finally he performs some evaluations over a set of use cases.
- Bohm et al. [1] developed the voiDgen software to analyze RDF graphs and output statistical data in VoID. They propose to compute additional kinds of class, property, and link-based partitions, e.g., sets of resources connected via specific predicates. Moreover, they show that distributed analyses of RDF large datasets are feasible by using a distributed algorithm (Map-Reduce).

Regarding the works related to dataset summarizaton we can say there are some existing efforts such as A-Box Summary [4] for efficient consistency checking, Zhang et al. [8] for summarising ontologies based on RDF sentence graphs, and Li et al. [6] for user-driven ontology summarisation. However, both help the understanding rather than the exploitation, which is usually task oriented.

3   RDF Summarisation: The Entity Description Pattern

Given an RDF graph, the summarisation task is to generate a condensed description which can facilitate data exploitations. Different from existing ontology summarisation work, we put special emphasises on identifying a special type of basic graph patterns in RDF data, which is suitable for data exploitation. The assumption of this special focus is that there exist such building blocks for revealing the constitution of an RDF dataset in a way which can not only help the understanding of the data but also is capable to guide RDF data exploitation. The rationale behind the assumption is that RDF data exploitation are

---

[2] https://github.com/cygri/make-void

usually based on graph patterns, e.g., SPARQL queries are based on BGP: basic graph patterns.

The main novelty of our summarisation approach is that it summarises an RDF graph by another much smaller graph structure based on atomic graph patterns. The linking structure in such summary graph can be utilised to significantly decrease search spaces in various data exploitation tasks e.g., query generation and query answering. Furthermore, statistic results of pattern instances are precomputed and attached to the summary, which can help both better understanding about the dataset and more efficient exploitation operations on it.

Specifically, in this paper, we propose one definition of such building blocks, i.e., *Entity Description Patterns* (EDPs for short), which is defined in Definition 1.

Definition 1. *(Entity Description Pattern) Given a resource e in an RDF graph G, the entity description pattern of e is $P_e = \{C, A, R, V\}$, in which C is the set of its classes, A is a set of its data valued properties, R is the set of its object properties, and V is the set of e's inverse properties.*

To get a more concise representation of an RDF graph, we define a merge operation on EDPs which can further condense the graph pattern result (c.f. Definition 2).

Definition 2. *(EDP Merge) Given a set of EDPs $\mathcal{P}$, let C be the set of all class components in $\mathcal{P}$ and let $G_{\mathcal{P}}(c_i)$ be a subset of $\mathcal{P}$ whose elements share the same class components c_i. Then,* merge *function can be defined as follows:*

$$Merge(\mathcal{P}) = \{(c_i, \bigcup_{P_i \in G_{\mathcal{P}(c_i)}} Attr(P_i), \bigcup_{P_i \in G_{\mathcal{P}(c_i)}} Rel(P_i), \bigcup_{P_i \in G_{\mathcal{P}(c_i)}} Rev(P_i))|c_i \in C\} \quad (1)$$

where

- *$Attr(P_i)$ denotes the attribute component of P_i;*
- *$Rel(P_i)$ denotes the relation component of P_i;*
- *$Rev(P_i)$ denotes the reverse relation component of P_i.*

The rationale behind this merge operation is that entities of the same type(s) might be viewed as a set of homogeneous things. Given this idea, we can define an EDP function of an RDF graph as Definition 3.

Definition 3. *(EDP of RDF Graph) Given an RDF graph G, its EDP function is defined by the following equation.*

$$EDP(G) = Merge(\bigcup_{e \in G} P_e) \quad (2)$$

EDP Graph. EDP function of an RDF graph results with a set of atomic graph patterns. Most data exploitation tasks can be decomposed into finding more complex graph patterns which can be composed by these EDPs. To this end, it would be more beneficial to know how EDPs are connected to each other in the original RDF graph. Such information can be useful not only in decreasing search spaces (e.g., in query generation) but also for guiding the exploitation (e.g., browsing or linkage). With regards to this consideration, we introduce *RDF data summarisation* as the notion of EDP graph (cf. Definition 4) for characterising the linking structures in the original RDF graph.

Definition 4. *(EDP Graph) Given an RDF graph G, its EDP graph is defined as follows*

$$\mathcal{G}_{EDP}(G) = \{< P_i, l, P_j > |\exists e_i \in E(P_i), \exists e_j \in E(P_j), < e_i, l, e_j > \in G, \\ P_i \in EDP(G), P_j \in EDP(G)\} \quad (3)$$

where $E(P_i)$ denotes the instances of EDP P_i. Specifically, if P_i is not merged EDP, $E(P_i)$ is the set of entities whose EDP is P_i; if P_i is a merged one, $E(P_i) = \cup_{P_k \in P} E(P_k)$, where P is the set of EDPs from which P_i is merged.

Annotated EDP Graph. EDP graphs are further annotated with statistic results. For each node e, it is annotated with a number which is the number of solutions to $Q(x) \leftarrow C_e(x)$. For each edge $l(C_e, C_f)$, there is a tuple of (n_1, n_2), whose elements are the numbers of solutions to $Q_1(x) \leftarrow C_e(x), l(x, y), C_f(y)$ and $Q_2(y) \leftarrow C_e(x), l(x, y), C_f(y)$ respectively.

4   Demos: The Summary Based Data Exploitations

To evaluate and demonstrate the effectiveness of our definition of data building blocks i.e., EDP, in data exploitation scenarios, we implemented an EDP based data exploitation system for three types of tasks i.e., gaining big picture and browsing, generating queries and enriching datasets.

The user interface is shown in Fig. 1 which contains three panels. The upper part is the *Dataset Selection Panel*, which displays the list of datasets in current demo system. To switch to another dataset, one can simply click on its name

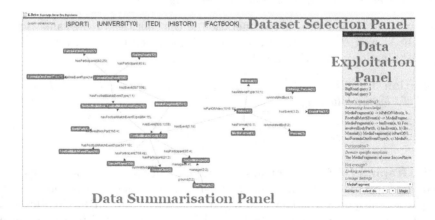

Fig. 1. Data Exploitation UI

in this panel. The middle panel is the main interaction and visualisation panel, the *Data Summarisation Panel*. By default, it displays the summarisation of the selected dataset as an interactive graph i.e., the EDP graph. In other situations, relevant subgraphs of the EDP graph will be shown in the data exploitation process. The right panel is the *Data Exploitation Panel*, which shows a bunch of UI components supporting various data exploitation operations.

Given the UI, we now demonstrate a list of data exploitation scenarios to illustrate how the summarisation can help the data exploitation tasks.

The Big Picture and Browsing Operations. When facing an unfamiliar dataset, users usually pursue a quick and rough *big picture* of it before (s)he can assess whether it is interesting or not, e.g., what are the data describing (concepts), how are the main concepts connected to each other (relations) and which are the important parts (clusters). To help the users gain answers to these questions quickly, as shown in the *Data Summarisation Panel* of Fig. 1, the EDP graph is visualised by using force-directed graph drawing techniques[3]. Each node in the graph describes a concept. In addition to the concept name, a node is also attached with the number of instances it has in the dataset. Such statistics(c.f. Figure 2) helps to assess the importance of each concept in the dataset (in terms of data portions). The relations between (instances of) these concepts are rendered as edges, and such edges are used to calculate groups of closely related nodes, which are in turn rendered as clusters in the graph. Two browsing operations are supported on the summary graph. The first is *node browsing*. By clicking on one node in the graph, users can gain detailed description about the concept (c.f. Figure 2) including the subgraph centralised on this node which is displayed in the middle panel and the natural language description of the node displayed in a pop-up panel on the left. The second browsing operation is *graph browsing*. After selecting a node, users can keep selecting/de-selecting interconnected nodes in current subgraph to grow or shrink it. This operation enables focused investigation on relations between interested nodes.

Query Generation. A typical usage on Semantic Web datasets is querying it. Query generation techniques [7] are helpful for either novice or advanced users because technical skills and dataset knowledge are prerequisites to write SPARQL queries. Based on the EDP summarisation, we implemented two types of query generation techniques. One is called guided query generation, which generates queries by utilising the EDP graph and statistics information attached in the graph. Such technique is good at generating queries for revealing main concepts and relations in the datasets. These two query types are called *Big City Queries* and *Big Road Queries* in the *Data Exploitation Panel* of the system. They are analogous to big cities and highways in a geography map. The other generation technique makes use of the links in the summarisation to do efficient

---

[3] Arbor Javascript Library (http://arborjs.org/introduction) is used for the EDP graph rendering.

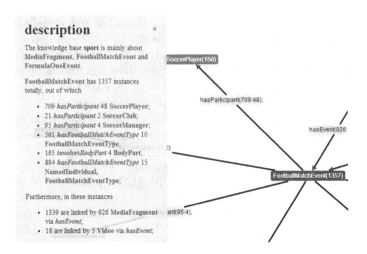

Fig. 2. Node Browsing

association rule mining [7]. This method is good at revealing insightful knowledge in the data in the form of corresponding graph patterns. Such queries are called *interesting knowledge* in the system. Clicking on any of these generated queries will bring out an illustrating subgraph in the middle part of the UI.

Dataset Enrichment. One of the promising features of Semantic Web techniques is the ability to link data silos to form a more valuable information space. Instead of instance-level linkage or ontology mapping, in our system, we introduce a new data linkage operation on EDPs. Such EDP-level linkage makes it possible to investigate what kinds of possibilities would be enabled after cross-dataset EDPs are linked, e.g., previously unanswerable queries might turn to be answerable by linking another dataset via EDP linkage. In the demo, we will demonstrate EDP-linkage between TED and Factbook datasets and show how such linkage can benefit a specific scenario of filtering tenders by country relations.

5 Conclusions and Future Work

We described a graph pattern based approach to facilitate RDF dataset exploitation. The rational behind this approach is to provide LEGO-like building blocks for users to play with RDF datasets. The building block proposed in this paper is so-called EDP, which is a basic entity description graph pattern. We have shown that (i) how EDPs can be constructed as an interactive summary to help quick understanding; (ii) how EDPs can be utilised to generate *interesting* queries under different interestingness definitions; (iii) how EDPs from different sources can be linked to each other so that unanswerable queries are turned to be answerable after this data enrichment. The future work will focus on investigating the properties of the summary and in-depth studies in above scenarios.

Acknowledgement. This research has been funded by the European Commission within the 7th Framework Programme/Maria Curie Industry-Academia Partnerships and Pathways schema/PEOPLE Work Programme 2011 project K-Drive number 286348 (cf. http://www.kdrive-project.eu). This work was also supported by NSFC with Grant No. 61105007 and by NUIST with Grant No. 20110429.

References

1. Bhm, C., Lorey, J., Naumann, F.: Creating void descriptions for web-scale data. Web Semant.: Sci., Serv. Agents World Wide Web **9**(3), 339–345 (2011)
2. Bizer, C., Heath, T., Berners-Lee, T.: Linked data-the story so far. Int. j. semant. web inf. syst. **5**(3), 1–22 (2009)
3. Auer, S., Demter, J., Martin, M., Lehmann, J.: LODStats – an extensible framework for high-performance dataset analytics. In: ten Teije, A., Völker, J., Handschuh, S., Stuckenschmidt, H., d'Acquin, M., Nikolov, A., Aussenac-Gilles, N., Hernandez, N. (eds.) EKAW 2012. LNCS, vol. 7603, pp. 353–362. Springer, Heidelberg (2012)
4. Fokoue, A., Kershenbaum, A., Ma, L., Schonberg, E., Srinivas, K.: The summary abox: cutting ontologies down to size. In: Cruz, I., Decker, S., Allemang, D., Preist, C., Schwabe, D., Mika, P., Uschold, M., Aroyo, L.M. (eds.) ISWC 2006. LNCS, vol. 4273, pp. 343–356. Springer, Heidelberg (2006)
5. Holst, T.: Structural analysis of unknown RDF datasets via SPARQL endpoints. Master thesis defense 11 (2013)
6. Li, N., Motta, E.: Evaluations of user-driven ontology summarization. In: Cimiano, P., Pinto, H.S. (eds.) EKAW 2010. LNCS, vol. 6317, pp. 544–553. Springer, Heidelberg (2010)
7. Pan, J.Z., Ren, Y., Wu, H., Zhu, M.: Query generation for semantic datasets. In: Proceedings of the seventh international conference on Knowledge capture, pp. 113–116. ACM (2013)
8. Zhang, X., Cheng, G., Qu, Y.: Ontology summarization based on rdf sentence graph. In: Williamson, C.L., Zurko, M.E., Patel-Schneider, P.F., Shenoy, P.J. (eds.) WWW, pp. 707–716. ACM (2007)

Zhishi.schema Explorer: A Platform for Exploring Chinese Linked Open Schema

Tianxing Wu[1(✉)], Guilin Qi[1], and Haofen Wang[2]

[1] Southeast University, Nanjing, China
{wutianxing,gqi}@seu.edu.cn
[2] East China University of Science and Technology, Shanghai, China
whfcarter@ecust.edu.cn

Abstract. Knowledge on schema level is vital for the development of Semantic Web, but the number of schema information in Linking Open Data (LOD) is limited. We approach this problem by contributing to the complementary part of LOD, that is, Linking Open Schema (LOS), which helps close the gap between lightweight LOD and expressive ontologies by adding more expressive ontological axioms between concepts. In this paper, we present Zhishi.schema Explorer for exploring Chinese linked open schema - Zhishi.schema. Zhishi.schema Explorer provides *Lookup Service* and *SPARQL Endpoint*, which respectively allow querying with concept labels and the SPARQL language.

Keywords: Linking Open Schema · Zhishi.schema · Zhishi.schema Explorer

1 Introduction

After the Linked Data [1] project initiated the efforts to connect the semantic data across the Web, there have been more than 200 datasets within the Linking Open Data (LOD)[1] cloud, which is the largest community effort for semantic data publishing. While LOD contains billions of triples describing millions of entities, their attributes and relationships, the number of schemas in current LOD is limited, let alone the schemas having labels in Chinese.

Yago [5,6] defines explicit schema to describe concept subsumptions as well as domains and ranges of properties. However, the quality of the schema is not always satisfactory. Freebase [3] has a very shallow taxonomy with domains and types. If we consider the schemas having labels in Chinese, the number is even smaller. DBpedia Ontology [2] (DBPO) enables users to define mapping rules to generate high-quality schema from ill-defined raw RDF data, but DBPO does not have the Chinese version. Zhishi.me [4] is the first effort to publish Chinese Linking Open Data, but it does not define an ontology to describe the schema information of the published semantic data.

---

[1] http://linkeddata.org/

© Springer-Verlag Berlin Heidelberg 2014
D. Zhao et al. (Eds.): CSWS 2014, CCIS 480, pp. 174–181, 2014.
DOI: 10.1007/978-3-662-45495-4_16

We approach the problem of schema sparseness by contributing to the complementary part of LOD, that is, Linking Open Schema (LOS). LOS aims at closing the gap between lightweight LOD and expressive ontologies by adding more expressive ontological axioms between concepts. Links in LOS are created between concepts from different sources and are not limited to equivalence relations. In this paper, we present Zhishi.schema Explorer, a platform for exploring Zhishi.schema [7], which is the first effort of publishing Chinese linked open schema. In the following, we first introduce the Zhishi.schema dataset in Sect. 2. Then, Sect. 3 describes Zhishi.schema Explorer, which provides *Lookup Service* and *SPARQL Endpoint* that allow querying with concept labels and the SPARQL language respectively. Finally, we conclude this paper and outline future work in Sect. 4.

2   Zhishi.schema

Zhishi.schema is not only an integrated concept taxonomy, but also a large semantic network. It consists of a directed graph where nodes represent concepts, and edges stand for semantic relations between them. These concepts and relations are harvested from navigational categories as well as dynamic tags in more than 50 various most popular Web sites in China, which cover all kinds of current Chinese social Web sites summarized in Fig. 1. Zhishi.schema comprises

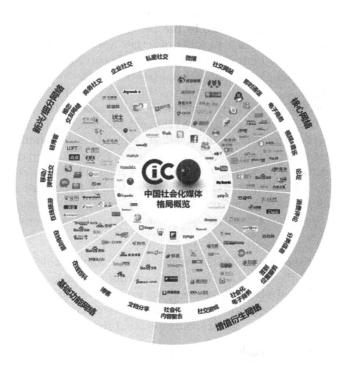

Fig. 1. The overview of current Chinese social Web sites

(a) Navigational (Static) Categories   (b) Dynamic Tags

Fig. 2. Typical examples of navigational (static) categories and dynamic tags

408,069 concept labels in which 328,288 are categories and 79,781 are tags, while the semantic relations include 1,560,725 `subclassOf` relations, 22,672 `equal` relations and 229,167 `relate` relations.

The navigational categories are organized in a hierarchical way. In a category hierarchy, a category might be associated with zero or several parent categories as well as child categories. A navigational category is called a static category as it is relatively stable and predefined by the Web site. The tags are organized in a flat manner and called dynamic tags because they are created on the fly by Web users. In fact, a tag can be treated as a single node category with no parents or children. Typical examples of navigational (static) categories and dynamic tags are given in Fig. 2.

According to the Linked Data principles[2], Zhishi.schema creates URIs for all concepts. The URI pattern `http://zhishi.schema/[site]/[concept type]/[label]` comprises of fours parts. http://zhishi.schema/ is the default namespace. The second part tells the provenance of the concept. For a category, the third part of its URI is static. If it is a tag, the part is dynamic. The last part is the label of a concept. In order to unify coding, the second and last parts are encoded into UTF-8[3]. Totally, Zhishi.schema contains 6 types of data: *labels, resource site labels, links, subclassOf relations, equal relations* and *relate relations*. They are explained in detail as follows:

– **Labels:** All concepts in Zhishi.schema have a name, which is used as a `rdfs:label` for the corresponding Zhishi.schema resource.
– **Resource Site Labels:** Since concepts are extracted from different Web sites, each of them has its own provenance. These provenances are represented with the predicate `zhishi.schema:resource_site_label`.

---

[2] http://www.w3.org/DesignIssues/LinkedData.html/
[3] http://www.utf-8.com/

- **Links:** For each concept, Zhishi.schema provides a link to allow users to access the original Web page. These links are extracted and represented using `zhishi.schema:site_url`.
- **SubclassOf Relations:** One concept is a subclass of another if and only if the former is a child node of the latter. In Zhishi.schema, `subclassOf` relations are denoted as `rdfs:subClassOf`.
- **Equal Relations:** Two concepts are equal if and only if they refer to the same meaning and this relation is represented using `owl:equivalentClass`.
- **Relate Relations:** Compared with `subclassOf` and `equal` relation, `relate` relation is the weakest semantic relation. Two concepts are related if their meanings are close but not the same. `skos:related` is used for representing `relate` relations.

In the Zhishi.schema dataset, all the `equal` and `relate` relations construct a large semantic network while all the `subclassOf` relations form an integrated concept taxonomy, which can be regarded as a hierarchical acyclic graph (*HAG*). The root depth is 1 and the maximal depth is 16. Since a concept may have one or more parents, we can traverse to the concept from the roots via different paths. These paths might have different lengths so that each concept could exist at multiple depths of *HAG*. On average, the depth of each concept is 3.479. The detailed information of concept (including static categories and dynamic tags) number and the number of and provenances for concepts at each depth of *HAG* is given in Table 1.

Table 1. Detailed information of HAG

Depth of HAG	Number of provenances for concepts	Concept number
1	51	18,925
2	51	14,280
3	49	97,997
4	43	95,342
5	43	37,423
6	38	18,986
7	36	9,725
8	35	6,684
9	28	5,148
10	17	3,287
11	12	2,026
12	8	589
13	6	152
14	4	44
15	2	17
16	1	6

3   Zhishi.schema Explorer

Zhishi.schema Explorer is a platform that allows users to explore the dataset of Zhishi.schema with *Lookup Service* and *SPARQL Endpoint*.

Lookup Service: Lookup service helps users to query with concept labels. It is available at http://los.linkingopenschema.info/LookUp.jsp and its interface is shown in Fig. 3. After submitting a user query, all concepts whose labels exactly match the query are returned. Since some of the concepts are equal, Zhishi.schema Explorer merges them and presents an integrated view for browsing. In contrast to the keyword search on the whole Web, query over the Zhishi.schema dataset can offer productive and useful knowledge on schema level directly rather than a large amount of texts or Web sites.

Fig. 3. The interface of lookup service

Figure 4 gives an example of the Lookup service. If one user searches for *"Water Purifier"*, a page which integrates two equivalent concepts from two e-commerce Web sites (360buy[4] and DangDang[5] respectively) is returned. The provenance information, other equivalent concepts, parent concepts, child concepts, related concepts, and links to Web pages of original Web sites are shown in the returned page. These information are organized in the `Resource Site Label`, `EqualClass`, `SuperClass`, `SubClass`, `RelatedClass` and `Link` section respectively.

Any parent concept or child concept can be clicked to switch to another page view. Such an interaction stands for navigation in the integrated concept taxonomy of Zhishi.schema. In addition, users can click on any related concept or equivalent concept and this interaction corresponds to traversal on the semantic network of Zhishi.schema.

---

[4] http://www.360buy.com/
[5] http://www.dangdang.com/

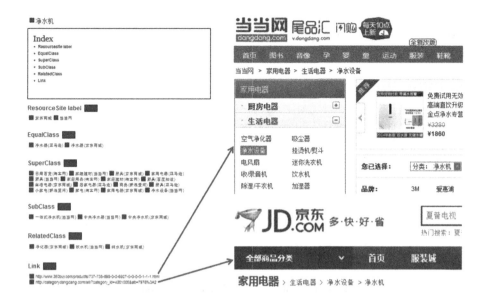

Fig. 4. An example of lookup service

SPARQL Endpoint: Zhishi.schema dataset can also be explored with a SPARQL endpoint, which is available at http://los.linkingopenschema.info/SPARQL.jsp. Figure 5 gives the interface of SPARQL Endpoint. This application is appropriate for the users who know in advance exactly what information is needed. These users can submit customized queries to this endpoint over the SPARQL protocol[6]. A query used to search for parent concepts of "*Water Purifier*" from DangDang is shown as follows:

SELECT ?product WHERE
{ < http://zhishi.schema/%E5%BD%93%E5%BD%93%E7%BD%91/static/
%E5%87%80%E6%B0%B4%E6%9C%BA >
< http://www.w3.org/2000/01/rdf − schema#subClassOf >
?product }

where UTF-8 in the URI of the subject represents encoded Chinese translations of "*DangDang*" and "*Water Purifier*". After submitting this query, 48 concepts are returned from Zhishi.schema. Table 2 gives part of the query results, including "*Water Purifying Plant*" from Amazon[7], "*Small Household Appliance*" from

---

[6] http://www.w3.org/TR/sparql11-protocol/

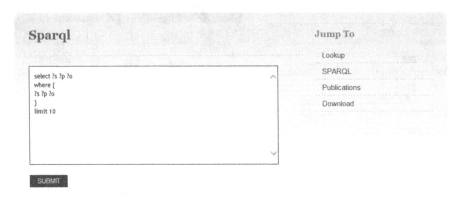

Fig. 5. The interface of SPARQL endpoint

Table 2. Sample query results from the SPAQRL endpoint

product
<http://zhishi.schema/ %E6%B7%98%E5%AE%9D%E7%BD%91(**Taobao**)/static/ %E5%AE%B6%E7%94%B5(**Household Appliance**)>
<http://zhishi.schema/ %E6%B7%98%E5%AE%9D%E7%BD%91(**Taobao**)/static/ %E5%B0%8F%E5%AE%B6%E7%94%B5(**Small Household Appliance**)>
<http://zhishi.schema/ %E5%BD%93%E5%BD%93%E7%BD%91(**DangDang**)/static/ %E5%AE%B6%E7%94%A8%E7%94%B5%E5%99%A8(**Household Appliance**)>
<http://zhishi.schema/ %E4%BA%9A%E9%A9%AC%E9%80%8A(**Amazon**)/static/ %E5%87%80%E6%B0%B4%E8%AE%BE%E5%A4%87(**Water Purifying Plant**)>

Text in brackets, which is English translation of decoded UTF-8, does not appear in the original query results.

Taobao[8] and "*Household Appliance*" from DangDang and Taobao. All the RDF triples are stored in AllegroGraph RDFStore[9] which also provides querying capabilities.

---

[7] http://www.amazon.cn/
[8] http://www.taobao.com/
[9] http://franz.com/agraph/allegrograph/

4   Conclusion

In this paper, we first introduced the Zhishi.schema dataset containing an integrated concept taxonomy with `subclassOf` relations and a large semantic network composed of `equal` relations as well as `relate` relations. More information concerning Zhishi.schema can be found in [7]. Then, we presented Zhishi.schema Explorer with *Lookup Service* and *SPARQL Endpoint*, a platform for exploring Zhishi.schema, which is the first effort of publishing Chinese linked open schema. It allows users to query with concept labels and the SPARQL language.

As for future work, we consider two aspects. First, Zhishi.schema will be linked to other datasets in LOD in order to build a global LOS. Second, we plan to provide more APIs as a programming interface of Zhishi.schema, thereby fostering more research with respect to analysing or mining the knowledge on schema level leveraging Zhishi.schema.

References

1. Bizer, C., Heath, T., Berners-Lee, T.: Linked data-the story so far. Int. j. semant. web inf. syst. **5**(3), 1–22 (2009)
2. Bizer, C., Lehmann, J., Kobilarov, G., Auer, S., Becker, C., Cyganiak, R., Hellmann, S.: DBpedia-a crystallization point for the web of data. Web Semant.: Sci., Serv. Agents world wide web **7**(3), 154–165 (2009)
3. Bollacker, K., Evans, C., Paritosh, P., Sturge, T., Taylor, J.: Freebase: a collaboratively created graph database for structuring human knowledge. In: Proceedings of the 2008 ACM SIGMOD international conference on Management of data (SIGMOD 2008), pp. 1247–1250 (2008)
4. Niu, X., Sun, X., Wang, H., Rong, S., Qi, G., Yu, Y.: Zhishi. me-weaving chinese linking open data. In: Proceedings of the 10th International Semantic Web Conference (ISWC 2011), pp. 205–220 (2011)
5. Suchanek, F.M., Kasneci, G., Weikum, G.: Yago: a core of semantic knowledge. In: Proceedings of the 16th international conference on World Wide Web (WWW 2007), pp. 697–706 (2007)
6. Suchanek, F.M., Kasneci, G., Weikum, G.: Yago: A large ontology from wikipedia and wordnet. Web Semant.: Sci., Serv. Agents World Wide Web **6**(3), 203–217 (2008)
7. Wang, H., Wu, T., Qi, G., Ruan, T.: On publishing Chinese linked open schema. In: Proceedings of the 13th International Semantic Web Conference (ISWC 2014), pp. 293–308 (2014)

Semantic Technology Applications

Evidence-Based Treatment
of Medical Guideline

Pingfang Tian[1,2], Zhonghua Zhu[1,2(✉)], and Zhisheng Huang[3]

[1] College of Computer Science and Technology, Wuhan University of Science
and Technology, Wuhan 430065, People's Republic of China
[2] Hubei Province Key Laboratory of Intelligent Information Processing
and Realtime Industrial System, Wuhan 430065, People's Republic of China
webartisan@yeah.net
[3] Department of Computer Science, Vrije University of Amsterdam,
1081 HV Amsterdam, The Netherlands

Abstract. Medical guidelines are recommendations on the appropriate treatment and care of people with specific diseases and conditions. Evidence-based medical guidelines are the document or recommendations which have been annotated with their relevant medical evidences, namely research findings from medical publications. We have observed the fact that there exist significant amount of medical guidelines have not yet annotated with relevant medical evidences, which becomes even more serious in the Chinese medical guidelines. In this paper, we propose an approach of evidence process of medical guidelines, such that we can find relevant evidences for those non-evidence-based medical guidelines. We develop a system called Link2Pubmed, which can retrieve the text which is described with a natural language and get the corresponding medical evidences. We use the word segmentation and part-of-speech tagging tools in natural language processing (NLP) to extract the keywords, and then translate them into corresponding English concepts in SNOMED CT, a well-known medical ontology. This system is an attempt to solve the existing problems in Chinese medical guidelines, which lack the annotations of relevant evidences.

Keywords: Medical guideline · NLP · SNOMED CT · PubMed · Link2Pubmed

1 Introduction

Clinical guidelines are the summary of the clinical practice work experience, the analysis of randomized controlled clinical study report and the consensus after discussion of recognized experts with high academic level. The guidelines have characteristics of openness and universality which can be used as a reference when clinical doctors deal with specified file [1] and clinical problems.

Clinical guideline is the bridge connecting the clinical evidence and clinical practice, reflects the status of the best clinical evidence. Due to its formulation process under specific racial, geographical, economic level, people's value and cultural factors,

D. Zhao et al. (Eds.): CSWS 2014, CCIS 480, pp. 185–197, 2014.
DOI: 10.1007/978-3-662-45495-4_17

using clinical research evidences in the native region is the most valuable. But influenced by medical conditions and the scientific research, the current Chinese medical communities are lack of a large number of clinical research evidences, especially the high quality randomized controlled study. So when making Chinese clinical guidelines, we often have to refer and use high quality clinical evidence abroad. In recent years, China has a certain amount of high quality clinical and randomized controlled clinical study [2]. However, the existing Chinese clinical guidelines are lack of the support of high quality evidences, which tend to be recognized by Chinese experts after discussion of consensus of high academic level, regarding as the recommendations.

For a clinical guideline with high quality, the most basic and most important issue is to make them evidence-based ones, including comprehensive collecting evidence and evidence for scientific and accurate evaluation.

Various academic groups in China encounter a lot of problems in the process of writing clinical diagnosis and treatment guidelines, the restricted factor is that the high quality randomized controlled clinical study is too little in China. Of course, we can't give up developing clinical guidelines because of a lack of the Chinese high level clinical research. Considering circumstances influenced by some factors, we can choose high-quality randomized controlled clinical studies in those countries which are also suitable for China. So, many Chinese scholars retrieve evidences on medical website abroad [3] (e.g. NGC [4], Cochrane Library [5, 6]) by manual, which is time consuming.

In this paper, we propose an approach of evidence process of medical guidelines, such that we can find relevant evidences for those non-evidence-based medical guidelines. We develop a system called Link2Pubmed, which can retrieve the text which is described with a natural language and get the corresponding medical evidences. We use the word segmentation and part-of-speech tagging tools in natural language processing (NLP) [7, 8] to extract the keywords, and then translate them into corresponding English concepts in SNOMED CT [9, 10], a well-known medical ontology. This system is an attempt to solve the existing problems in Chinese medical guidelines, which lack relevant evidences. Our initial experiments show that our approach is efficient for the target.

The rest of this paper is organized as follows: Sect. 2 presents the general idea of Link2Pubmed. Section 3 shows the system processing flow of Link2Pubmed in detail. Section 4 presents the design and implementation of the system module. Section 5 shows the implementation details, system test and evaluation. Section 6 discusses the related work, future work, and makes the conclusions.

2 Link2Pubmed System

For the situation that many Chinese clinical guidelines lack annotated with relevant medical evidences, this paper proposes a solution to evidence-based treatment of existing medical guidelines, namely according to the fact guidelines described to find corresponding medical evidence. Due to the current status of lacking evidence in Chinese clinical guidelines, our approach is to retrieve evidences on PubMed, and converts the data obtained from PubMed into ones with required formats, and finally present them in the user interface of the system.

PubMed [11] is a free search engine, providing search and a summary of biological medicine. Its database source is MEDLINE. Its central theme is medicine, but also includes other medical field related to it, such as nursing or other health disciplines. It also provides support for the related biomedical information on the quite comprehensive, such as biochemistry and cell biology. Journal articles in free PubMed information service do not include the full text, but may provide links to the provider. MEDLINE collected articles from 1966 until now, including medical, nursing, veterinary medicine, health care system and literature of the preclinical science article. These data from more than 70 countries and regions, more than 4800 biomedical journals, in recent years, the data involved in more than 30 languages, dating back to 1966 years of data involved in more than 40 languages, around 90 % of English literature, 70 % ~ 80 % of literature with the author to write English abstract. So the retrieved medical evidence on PubMed can guarantee finding enough medical evidence.

However, retrieval on PubMed needs to provide medical subject headings which will be retrieved, so for the guidelines described in natural language text, they must be transformed from Chinese natural language description guideline to medical subject headings which can be retrieved on PubMed. This paper designs and implements the system, Link2Pubmed. From the perspective of guideline text described in Chinese natural language, this paper extract keywords from guideline text, then converted them into corresponding medical subject headings, then go on PubMed search. In the third section of the article tells the whole system design process.

3 Link2Pubmed System Processing Flow

The process of Link2Pubmed system design is shown in Fig. 1. The figure described the whole system of the Link2Pubmed data processing flowchart.

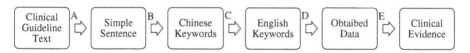

Fig. 1. Link2Pubmed system flowchart

In the whole system, there are 6 kinds of forms of data: Clinical Guidelines Text, Simple Sentence, Chinese Keywords, English Keywords, Obtained Data, Clinical Evidence. There are 5 kinds of options between the 6 kinds of forms data, which are:

A. Semi-automatic process for clinical guidelines. Through semi-automatic processing, Clinical guideline texts generate the corresponding simple sentence patterns of text;
B. Extracting keywords. For simple sentence text, keyword extraction can get the Chinese keywords;
C. Keywords conversion from Chinese to English. In order to retrieve the keywords extracted from guideline text in PubMed search, we must transform Chinese keywords into the corresponding English medical concepts.

D. PubMed retrieval. Retrieving in PubMed using translated English keywords can get the corresponding medical evidence.

E. Data formatting. The data retrieved from the PubMed contains large amounts of information, the information expressed in XML format, in order to make the user more intuitive to see the data; we must manipulate the data format in another way.

3.1 Semi-Automatic Process for Clinical Guidelines

Because of the limitation of the current natural language processing technology, it can't implement word segmentation and part-of-speech tagging [12, 13] with a whole guideline or a large section of a guideline. This is because with the increase of processing text length, the consumed time that the word segmentation tool spend to deal with text will have exponentially times growth, at the same time, the accuracy of the result of word segmentation and part-of-speech tagging will be greatly reduced. So when to extract keywords in the guideline, the object to deal with should be a simple sentence, rather than a large section of the text. So, transforming guideline text into simple sentences after word segmentation and part-of-speech tagging is a good solution. But medical guidelines which are described by natural language can't simply use punctuation marks to divide the text to achieve the purpose of transforming guideline text into a simple sentence because of the syntactic structure, grammatical structure and other logical relationship of natural language. There is no such tools can smartly implement text segmentation with a description of natural language. So we can only use a semi-automatic method, namely using artificial methods in natural language processing, analyzing the text structure and the logical relationship between sentences guide and for text segmentation.

3.2 Extracting Keywords

For the simple sentence after semi-automatic processing, keywords extracted from them involved in symptoms, disease, medicines and medical terms, etc. We can know what the meaning of the sentence with these words is. In this design, the word segmentation tool is ICTCLAS (Institute of Computing Technology, Chinese Lexical Analysis System) [14] which comes from the Chinese academy of sciences. The main features include: Chinese word segmentation, the part of speech tagging, named entity recognition; new word recognition, etc., at the same time, it also supports custom glossary, which greatly help to extract keywords and phrases in guidelines like symptoms, disease, medicines and medical terms, etc.

3.3 Keywords Conversion from Chinese to English

After getting the keywords, we can't retrieve them before translation. Using some medical dictionary for translation, it can be done for key words or phrases expressed meaning transformation in both English and Chinese, but it often doesn't present well for these words and phrases have corresponding medical concepts. SNOMED CT (Systematized Nomenclature of Medicine, Clinical Terms), is a system organized, an advantageous set of the medical term for computer processing, covering most aspects of Clinical information, such as disease, can see, the operation and microorganism, drugs, etc. Using the term set can be coordinated between different disciplines and specialties, and the location of the care for clinical data indexing, storage, retrieval, and aggregation. This greatly improves the accuracy of the translation.

3.4   PubMed Retrieval

PubMed offers open data query link, the user can send the request directly to call this interface. What we need to do is using English keywords to generate a request URL, and then sends an HTTP request to the PubMed, it will return the relevant data and information.

3.5   Data Formatting

Information retrieved from a PubMed is expressed in XML format which store huge amounts of information and only a few of this information is really needed to users, which allows the user effectively get useful information. In order to present the retrieved data more clearly to the user, the system will output data in a form that a user looks more intuitive after parsing XML.

4   System Module Design and Implementation

Based on the system processing flow, Link2Pubmed system module design is shown in Fig. 2:

4.1   Guideline Semi-Automatic Processing

To deal with the text of the guideline described in Chinese, before and after the Chinese syntactic structure and logic is relatively complex, so in this system the semi-automatic method is adopted to process guideline text. But in the process of dealing with guideline text, the following points must comply with:

First, must make the pledge that the meaning of the sentence is as same with the original text. Namely, in the process of processing, convert guideline text equivalent into simple sentences. This is one of the most important in the process of guide semi-automatic processing.

Second, pay attention to the logical relation of the sentence with the other before and after it. In the guideline described in Chinese, due to the logic relationship with the sentence before and after it, sometimes a few words before and after the same aspect of description are presenting the same truth, so we should combine several sentences into one sentence, describing a fact.

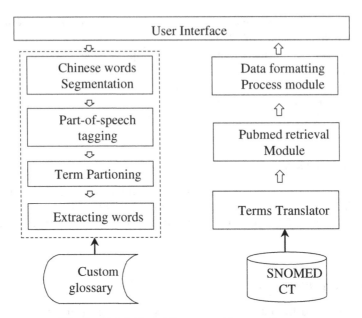

Fig. 2. Link2Pubmed System architecture diagram

Third, pay attention to the hierarchical relationships in a single sentence. In the Chinese guideline, a sentence can have more clauses, describing various aspects of a fact, the amount should be subdivided each clause, also should be a sentence to describe a fact.

The sentences after semi-automatic processing should be clear, eventually a sentence to describe a fact which has no superfluous information and doesn't miss any information.

Figure 3 is a text taken from Chinese antimicrobial treatment guidelines, which describes what rules should pay attention to when using the penicillin. In this text, the sentences are combinations of complex sentences, which describe the various aspects of a fact in one sentence, just the fact that it can be subdivided into several facts, namely one sentence describes only fact. Figure 4 is the result of semi-automatic processing, the original of the complex sentence is broken down into several simple sentences, and each simple sentence describes only one truth.

　　1. 无论采用何用种给药途径，用青霉素类药物前必须详细询问患者有无青霉素类过敏史、其他药物过敏史及过敏性疾病史，并须先做青霉素皮肤试验。

　　2. 过敏性休克一旦发生，必须就地抢救，并立即给病人注射肾上腺素，并给予吸氧、应用升压药、肾上腺皮质激素等抗休克治疗。

　　3. 全身应用大剂量青霉素可引起腱反射增强、肌肉痉挛、抽搐、昏迷等中枢神经系统反应（青霉素脑病），此反应易出现于老年和肾功能减退患者。

Fig. 3. Segment before semi-automatic processing

用青霉素药物前，必须询问患者有无青霉素类过敏史。
用青霉素药物前，必须询问患者有无其他药物过敏史。
用青霉素药物前，必须询问患者过敏性疾病史。
用青霉素药物前，必须先做青霉素皮肤试验。
过敏性休克一旦发生，必须就地抢救。
过敏性休克一旦发生，应立即给病人注射肾上腺素。
过敏性休克一旦发生，应立即给病人吸氧。
过敏性休克一旦发生，应用升压药抗休克治疗。
过敏性休克一旦发生，应用肾上腺皮质激素抗休克治疗。
全身应用大剂量青霉素可以起腱反射增强。
全身应用大剂量青霉素可以起肌肉痉挛。
全身应用大剂量青霉素可以起抽搐。
全身应用大剂量青霉素可以起昏迷。

Fig. 4. Segment after semi-automatic processing

4.2   Keyword Extraction Module

Custom glossary. To extract reasonable keywords from a sentence, the first thing needs to know is which words should be extracted. ICTCLAS is a tool that can customize the word in the glossary, and in this glossary we can specific a part of speech for a specified custom word, for example, "penicillin" can specify part of speech of "YP", namely "drug"; We can also specify the "fever" as "ZZ", namely the symptoms. Custom glossary is a large library, and all of the keywords to be extracted are included in the library. The repository contains all kinds of drugs, symptoms, disease, and qualitative values and other medical terms. The complement of the custom glossary determines the keyword extraction accuracy.

Word segmentation and part-of-speech tagging. Using ICTCLAS, sentences can be divided into individual words, separated by spaces, at the same time for each term labeling part of speech, so that you can pick out the word with an annotation according to the specific part of speech, which are the keywords to be extracted [15]. ICTCLAS is a good tool, in this operation, what we need to do is to load the custom glossary and invoke the ICTCLAS segmentation and part-of-speech tagging interface.

Word segmentation and extraction. After word segmentation and part-of-speech tagging, sentence is still in the form of a string, but we can use the characteristics of the string after marking to implement segmentation. After word segmentation and part-of-speech tagging, we can get a combination of words and part of speech, and there is a space separated between words and words, this is really the whole sentence can be divided into individual words, choose the custom part of speech of words, which is in the end we want to extract.

Translation Module. We use SNOMED CT, the well-known medical ontology, for the translation support in the translation module. The main advantage to use a medical ontology for the translation process is that it provides a standard terminology set for the medical domain. We obtain the standard set of the medical concepts with their English labels and use the Google translation to obtain their corresponding Chinese terms [16].

We design a local user translation dictionary with both English and Chinese terms, which consist of concept ID in SNOMED CT, type of the term, such as YP, which stands for drug, and others, the English term, and the Chinese term, like this:

```
61651006|YP|Cefamandole|头孢孟多
61651006|YP|Cefamandole|头孢孟多(物质)
61651006|YP|Cefamandole|头孢孟多(产品)
61651006|YP|Cefamandole|羟苄四唑头孢菌素
61651006|YP|Cefamandole (substance) |头孢孟多
61651006|YP|Cefamandole (substance) |头孢孟多(物质)
61651006|YP|Cefamandole (substance) |头孢孟多(产品)
61651006|YP|Cefamandole (substance) |羟苄四唑头孢菌素
61651006|YP|Cefamandole (product) |头孢孟多
61651006|YP|Cefamandole (product) |头孢孟多(物质)
61651006|YP|Cefamandole (product) |头孢孟多(产品)
61651006|YP|Cefamandole (product) |羟苄四唑头孢菌素
61651006|YP|Cephamandole|头孢孟多
61651006|YP|Cephamandole|头孢孟多(物质)
61651006|YP|Cephamandole|头孢孟多(产品)
61651006|YP|Cephamandole|羟苄四唑头孢菌素
61862008|YP|Methicillin|耐甲氧西林
61862008|YP|Methicillin|耐甲氧西林(物质)
61862008|YP|Methicillin|甲氧西林(产品)
```

Although there might exist multiple matching in the translation dictionary, we would always prefer a translated term which is shortened to others. For those which cannot be translated by using the local translation dictionary, we use the Baidu translator to do the complement work.

4.3   PubMed Retrieval Module

Translating module will output English keywords. If we want to get the PubMed IDs (PMIDs) for articles about breast cancer. The query string can be written as follows: http://eutils.ncbi.nlm.nih.gov/entrez/eutils/esearch.fcgi?db=pubmed&term=breast +cancer. If you want to know more about Entrez Programming, you can log on http:// www.ncbi.nlm.nih.gov/books/NBK25500/ for more information.

4.4   Data Formatting Module

The XML parser is used to convert the parse XML data into a particular data structure. It only stores some fundamental information such as title, authors, PMID, journal, abstract, published time. Link2Pubmed use a structure array to store these properties which have been formatted.

5   Implementation, Test and Evaluation

5.1   System Implementation

After starting Link2Pubmed, we will see the interface as shown in Fig. 5. The system interface is very simple, only one input box and one output. But when we input a Chinese sentence after semi-automatic processing in the input box, it will obtain the related medical evidence after retrieval.

Through semi-automatic processing, we get the Chinese guideline text composed of some simple sentences. For example, there is now a simple sentence after semi-automatic processing: "如果患者是妊娠期患者,那么应避免使用甲硝唑" (If the patient is a patient with pregnancy, so avoid using metronidazole.). Input this simple sentence in the "Source" Column of Link2Pubmed system. Through the analysis of the Link2Pubmed, it can extract two Chinese keywords: "妊娠" (pregnancy) and "甲硝唑" (metronidazole), and it can convert to the corresponding English medical terms "Pregnancy" and "Metronidazole". After retrieval, we can get the results shown in Fig. 5. In Fig. 5, we can see some medical evidence in the results section lists, and each piece of evidence is given some basic information such as title, author, journal name

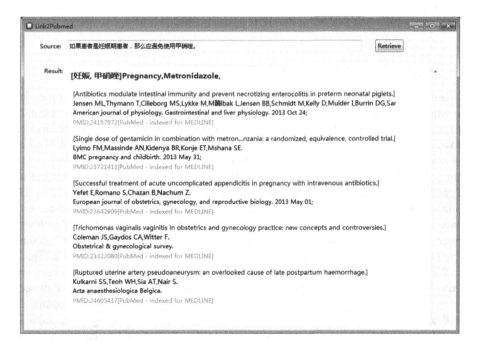

Fig. 5. Link2Pubmed Retrieval Result

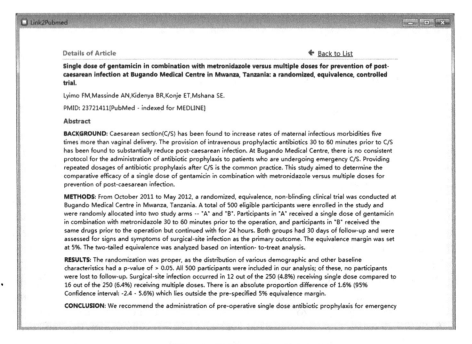

Fig. 6. Evidence Details

and PMID. Click on the title of each piece of evidence to view the details of this evidence, as shown in Fig. 6. In Fig. 6, it shows the abstract of the evidence in detail.

5.2   Experiment and Evaluation

We use the Chinese clinical guidelines for rational use of antibiotics as the test data for the experiment. We select 100 statements after semi-automatic processing as test data of Link2Pubmed randomly. We can define a stipulated as follows: KA1 is the amount of keywords in the sentence, KA2 is the amount of keywords which be found, KA3 is the amount of keywords in the sentence which exist in custom glossary, FP1 is the percentage of keywords being found actually (KA2/KA1), FP2 is the percentage of keywords being found in Link2Pubmed with custom glossary (KA2/KA3), TP is the accuracy completion percentage of translation, and IA shows whether or not the evidence is available. We calculate the average value of FP1, FP2, TP and IA of each test sentence Meanwhile, we set two groups of data which are respectively with uncompleted custom glossary and completed custom glossary. The results are shown in Fig. 7.

We can see the item FP1 in Fig. 7, as long as the keywords of test sentences are already defined in the custom glossary the words will be found out in Link2Pubmed, but in fact the custom glossary is uncompleted, the number of words didn't contain all medical vocabulary, which makes that the actual keyword extracting percentage is 76.7 %, not 100 %. At the same time, we should pay attention to that the average IA, which means whether the evidence is available, is only 0.4. This is directly related

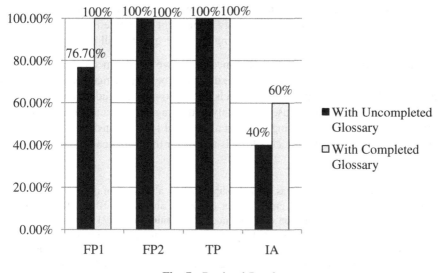

Fig. 7. Retrieval Result

to the accuracy and completeness of keyword extracting. With the completed custom glossary of test data, the actual keyword extracting percentage is 100 %, which also improves the availability of the evidences in a degree, this mainly makes up for the factors that the availability of evidences is not high caused by incomplete keyword exacting. One another factor determining the availability of evidences is the accuracy of keyword translation; on the other hand, there is indeed no corresponding evidence in PubMed. However, the system can be improved if a high-quality custom glossary is provided.

6 Conclusion and Future Work

In order to solve the problem that the Chinese clinical evidence is insufficient, Link2Pubmed provides an approach to process the existing medical guidelines for evidence-based treatment. Link2Pubmed can retrieve guideline text, and automatically extract keywords, and convert them to the corresponding medical concepts, and retrieve on PubMed accurately and effectively, so as to get a lot of medical evidences.

Link2PubMed provides a tool for Chinese researchers for medical evidence retrieval. Here are some future works we are going to improve the system:

1. The keywords are not able to be maximized extracted. Keyword extraction mainly depends on the custom glossary. So, a glossary which covers more medical terms will make the extraction of keywords more completely.
2. The keywords between English ones and Chinese ones are not totally transform-able. Some extracted Chinese keywords may not find their corresponding concepts in SNOMED CT. Under that situation, we have to use some additional dictionaries to translate them into corresponding English terms. The cost of guidelines'

semi-automatic processing is high. Depending on the current level of natural language processing, we cannot reasonably implement the guideline text conversion, but with the development of natural language, the quality will be improved.

One of the main features of Link2Pubmed is that we use the medical ontology such as SNOMED CT in our system. As we have discussed before, the main advantage to use SNOMED CT in our Link2Pubmed system is that it provides a standard terminology set in the medical domain. That is useful not only in the fragmentation and POS processing in the natural language processing tool, but also useful for the translation process. That shows that the Semantic Web technology is useful for the evidence process of medical guidelines.

Acknowledgement. This work was partially supported by a grant from the NSF (Natural Science Foundation) of China under grant number 60803160 and 61272110, the Key Projects of National Social Science Foundation of China under grant number 11&ZD189, and it was partially supported by a grant from NSF of Hubei Prov. of China under grant number 2013CFB334. It was partially supported by NSF of educational agency of Hubei Prov. under grant number Q20101110, and the State Key Lab of Software Engineering Open Foundation of Wuhan University under grant number SKLSE2012-09-07.

References

1. Xiao-wei, J., Zhu-ming, J.: The present status of Chinese clinical practice guidelines. Chin. J. Clin. Nutr. **18**(6), 327–329 (2010)
2. Jiang, Z.M., Wilmore, D.W., Wang, X.R., et al.: Randomized clinical trial of intravenous soybean oil alone versus soybean oil plus fish oil emulsion after gastrointestinal cancer surgery. Br. J. Surg. **97**(6), 804–809 (2010)
3. He, L., Du, X.L.: Retrieval research of evidence based clinical practice guidelines. J. Clin. Rehabilitative Tissue Eng. Res. **11**(40), 8173–8177 (2007)
4. Mast, K.R., Salama, M., Silverman, G.K., Arnold, R.M.: End-of-life content in treatment guidelines for life-limiting diseases. J. Palliat. Med. **7**(6), 754–773 (2004)
5. Kennedy, S., Bergqvist, A., Chapron, C., D'Hogghe, T., et al.: ESHRE guideline for the diagnosis and treatment of endometriosis. Hum. Reprod. **20**(6), 2698–2704 (2005)
6. Burton, M.J., Couch, M.E.: Rosenfeld. R.M.: Extracts from the cochrane library homeopathic medicines for adverse effects of cancer treatments. Otolaryngol. Head Neck Surg. **141**(2), 162–165 (2009)
7. Collobert, R., Weston, J.: A unified architecture for natural language processing: deep neural networks with multitask learning. In: Proceeding of the 25th International Conference on Machine, pp. 160–167 (2008)
8. Chopra, A., Prashar, A., Sain, C.: Natural language processing. Int. J. Enhancements Emerg. Eng. Res. **1**(4), 131–134 (2013)
9. Elkin, P.L., Brown, S.H.: Evaluation of the content coverage of SNOMED CT: ability of SNOMED clinical terms to represent clinical problem lists. Mayo Clin. Proc. **81**(6), 741–748 (2006)
10. Wasserman, H., Wang, J.: An applied evaluation of SNOMED CT as a clinical vocabulary for the computerized diagnosis and problem list. AMIA Annu. Symp. Proc. **2003**, 699–703 (2003)

11. 沈锡宾, 吕小东, 郝秀原, 孙静, 汪谋岳, 郭利劭. PubMed Central简介及其对期刊的评估和收录. 中国科技期刊研究 **17**(5), 866–868 (2006) (XiBin, S., XiaoDong, L., XiuYuan, H., Jing, S., MouYue, W., LiShao, G.: PubMed central introduction and a review and record of journal. Chin. J. Sci. Tech. Periodicals **17**(5), 866–868 (2006))

12. Jiang, W., Huang, L., Liu, Q., Lü, Y.: A cascaded linear model for joint chinese word segmengation and part-of-speech tagging. In: Proceeding of ACL, pp. 897–904 (2008)

13. Jiang, W., Mi, H., Liu, Q.: Word lattice reranking for Chinese word segmentation and part-of-speech tagging. In: Proceedings of the 22nd International Conference on Computational Linguistics, pp. 385–392 (2008)

14. Zhang, H., Yu, H., Xiong, D., Liu, Q.: HHMM-based Chinese lexical analyzer ICTCLAS. SIGHAN '03 Proceedings of the Second SIGHAN Workshop on Chinese Language Processing, pp. 184–187 (2003)

15. 夏天, 樊孝忠, 刘林. 利用JNI实现ICTCLAS系统的Java调用. 计算机应用 **24**(Z2), 177–178 (2004). (Xia, T., Fan, X., Liu, L.: Java calls using JNI ICTCLAS system. Comput. Appl. **24**(Z2), 177–178 (2004))

16. Ceusters, W., Smith, B., Kumar, A.: Dhaen, A.: Ontology-based error detection in SNOMED-CT. In: Proceedings of MEDINFO 482–486 (2004)

Evidence-Based Clinical Guidelines in SemanticCT

Qing Hu[1,2](✉), Zhisheng Huang[1], Frank van Harmelen[1], Annette ten Teije[1], and Jinguang Gu[2]

[1] Department of Computer Science, VU University Amsterdam,
Amsterdam, The Netherlands
{huang,Frank.van.Harmelen,annette}@cs.vu.nl
[2] Faculty of Computer Science and Engineering,
Wuhan University of Science and Technology, Wuhan, China
{qinghu,simon}@ontoweb.wust.edu.cn

Abstract. Evidence-based Clinical Guidelines are the document or recommendation which follow a rigorous development process and are based on the highest quality scientific evidence. Evidence-based clinical guidelines are important knowledge resources which have been used in many medical decision support systems and medical applications. In this paper, we present a semantic approach of evidence-based clinical guidelines. That lightweight formalisation of clinical guidelines have been integrated with SemanticCT, a semantically-enabled system for clinical trials. We have developed several tools to generate semantic data from textual guidelines. We show how they are useful in SemanticCT for the applications of the Semantic Web technology in medical domains.

1 Introduction

Clinical guidelines are recommendations on the appropriate treatment and care of people with specific diseases and conditions. Clinical guidelines have been proved to be valuable for clinicians, nurses, and other healthcare professionals in their work. Evidence-based Clinical Guidelines are that the document or recommendation which have been annotated with the best medical findings. Those medical findings can be found from medical publications, like PubMed.

Evidence-based clinical guidelines are important knowledge resources which have been used in many medical decision support systems and medical applications. However, existing evidence-based clinical guidelines are usually in the textual format. They are not ready for reuse in a medical knowledge-based system or a decision making system. There exist some formalisms for Computerized Clinical Guidelines, alternatively called Computer-Interpretable Guidelines (CIGs), like those in PROforma [4,5], Asbru [15], EON [16], GLIF [13,14], which implement the guidelines in computer-based decision support systems. They usually require manual generation of the formalization and have not yet supported for the semantic inter-operability.

© Springer-Verlag Berlin Heidelberg 2014
D. Zhao et al. (Eds.): CSWS 2014, CCIS 480, pp. 198–212, 2014.
DOI: 10.1007/978-3-662-45495-4_18

The semantic enriched guidelines are useful for several tasks:

- Browsing through the guidelines (for instance for evidence of recommendations);
- Support for the guideline update process: identifying relevant literature for a particular recommendation (or guidelines);
- Search for relevant clinical trials for a particular guideline/recommendation.

Due to large number of guidelines, those processes above should be done automatically or semi-automatically. In [8], we have proposed a semantic approach of evidence-based clinical guidelines, in which evidence-based clinical guidelines are represented as the format RDF/RDFS/OWL, the standards in the Semantic Web technology. We have used the NLP techniques to identify concepts and relations and developed several tools to generate semantic data from textual guidelines. In this paper, we will report how this approach of the semantics enriched formalization is integrated with SemanticCT, a semantically-enabled system for clinical trials [7]. We have implemented the component of evidence-based guidelines in SemanticCT. This guideline component in SemanticCT provides the functionality which integrates clinical guidelines with clinical trials. We will show how the semantic representation of evidence-based clinical guidelines are useful for the applications of the Semantic Web technology in medical domains.

The main contribution of this paper is:

1. presenting the design of the component of evidence-based clinical guidelines for SemanticCT,
2. showing how the semantic approach of evidence-based clinical guidelines is integrated with SemanticCT,
3. discussing several use cases of evidence-based clinical guidelines in the system.

This paper is organized as follows: Sect. 2 introduces the general ideas about semantic representation of evidence-based clinical guidelines. Section 3 presents SemanticCT, the semantically-enabled system for clinical trials and discusses the design of the guideline component in the system. Section 4 describes the implementation of the tools for evidence-based clinical guidelines and show how they can be used to convert the textual clinical guidelines into semantic data. Section 5 discusses several use cases on the evidence-based clinical guideline component in SemanticCT. Section 6 discusses future work and make the conclusions.

2 Semantic Representation of Evidence-Based Clinical Guidelines

Evidence-based Clinical Guidelines are a series of recommendations on clinical care, supported by the best available evidence in the clinical literature. In evidence-based clinical guidelines, the answers to the fundamental questions are based on published scientific research findings. Those findings are usually found in medical publications such as those in PubMed. The articles selected are evaluated by an expert in methodology for their research quality, and graded in proportion to evidence using the following classification system [12].

A classification of research results are proposed on level of evidence in [11,12] consists of the following five classes: (i) A1: Systematic reviews(i.e. research on the effects of diagnostics on clinical outcomes in a prospectively monitored, well-defined patient group), or that comprise at least several A2 quality trials whose results are consistent; (ii) A2: High-quality randomized comparative clinical trials (randomized, double-blind controlled trials) of sufficient size and consistency; (iii) B. Randomized clinical trials of moderate quality or insufficient size, or other comparative trials (non-randomized, comparative cohort study, patient control study); (iv) C: Non-comparative trials, and (v) D: Opinions of experts, such as project group members.

Based on the classification of evidences, we can classify the conclusions in the guidelines, alternatively called *guideline items*, with an evidence level. The following evidence levels on guideline items, are proposed in [11]: (i) Level 1: Based on 1 systematic review (A1) or at least 2 independent A2 reviews; (ii) Level 2: Based on at least 2 independent B reviews; (iii) Level 3: Based on 1 level A2 of B research, or any level C research, and (iv) Level 4: Opinions of experts.

Here is an example of the conclusion in an evidence-based clinical guidelines in [11]:

```
Level 1
The diagnostic reliability of ultrasound with an uncomplicated cyst is very high.
A1 Kerlikowske 2003
B Boerner 1999, Thurfjell 2002, Vargas 2004
```

which consists of an evidence level, a guideline statement, and their references with the classification of the research results.

In [8], we have proposed an approach of semantic representation for evidence-based clinical guidelines. We have designed RDF/OWL-based terminologies (thus a lightweight ontology) to express clinical evidences, so that those concepts can be used to represent various evidence information in clinical guidelines.

The semantic representation of evidence-based clinical guidelines consists of the following sections: (i) Heading. The heading section of the guidelines provide the basic description of the information such as the title and provenance; (ii) Body. The body section provides the main description of guidelines and their evidences. It consists of a list of guideline items contain the evidence information and their semantic representations of a single guideline statement.

Those evidence-based guidelines are represented by using the RDF/RDFS/OWL standards. Compared with existing other formalisms of Computer-Interpretable Guidelines, which usually require a lot of manual processing for the generation of the formulation, our lightweight formalism of evidence-based clinical guidelines has the advantage that they can be generated quite easy by using the tools have been developed in the NLP and the Semantic Web. We have shown that this semantic representation of evidence-based clinical guidelines have some novel features, and can be used for various application scenarios in the medical domain [8].

Here is an example of the semantic representation of evidence-based clinical guidelines in the RDF N-Triple format:

For each guideline item, we have the following statements of the evidence description, which are represented as follows:

```
sctid:gl002-zsh140412 sct:hasConclusions sctid:gl002-zsh140412_32.
sctid:gl002-zsh140412_32 rdf:type sct:GuidelineConclusion.
sctid:gl002-zsh140412_32 sct:about "Prognostic factors".
sctid:gl002-zsh140412_32 sct:hasGuidelineItem sctid:gl002-zsh140412_32_2.
sctid:gl002-zsh140412_32_2 sct:hasText "There are indications that
the presence of micrometastases or isolated tumour cells has a negative
influence on the disease-free survival with a hazard ratio of 1.5 at a
median follow-up of 5 years and that the relative risk reduction in relation
to the disease-free survival with adjuvant systemic treatment in this group
of patients is no different than in the general breast cancer population.".
sctid:gl002-zsh140412_32_2 sct:evidenceLevel "3"^^xsd:decimal.
sctid:gl002-zsh140412_32_2 sct:hasReferences sctid:gl002-zsh140412_32_2ref.
sctid:gl002-zsh140412_32_2ref sct:hasReference sctid:gl002-zsh140412_32_2ref1.
sctid:gl002-zsh140412_32_2ref1 sct:reference sctid:gl002-zshref-deBoer2009.
sctid:gl002-zsh140412_32_2ref1 sct:evidenceClassification "B".
sctid:gl002-zsh140412_32_2 sct:hasRelations sctid:gl002-zsh140412_32_2e8.
```

The statements above describe the guideline items with their evidence levels and references. A pointer 'sctid:gl002-zsh140412_32_2e8' is used to point to its initial node of the relations extracted by the NLP tool.

The NLP tool that we use is XMedlan, the Xerox linguistic-based module of the relation extraction system of the EURECA project [1]. The main characteristic of this component is that it uses a linguistic parser [2] to perform rich linguistic analysis of the input text. All the annotations produced by the linguistic analyzer are exploited by the relation extraction engine to identify relations and attributes of concepts and entities in the input text. These relations and attributes are expressed as triples, i.e. typed binary relations in the form ⟨Subject, Property, Object⟩. They are serialized in the RDF N-Triple format, like this:

```
sctid:gl002-zsh140412_32_2e1 ctec:hasUnit "years".
sctid:gl002-zsh140412_32_2e1 ctec:hasQuant "5".
sctid:gl002-zsh140412_32_2e2 ctec:hasQuant "1.5".
sctid:gl002-zsh140412_32_2e3 ctec:isA sct:diagnosis.
sctid:gl002-zsh140412_32_2e3 ctec:hasObject sctid:gl002-zsh140412_32_2e4.
sctid:gl002-zsh140412_32_2e4 ctec:hasTerm "breast cancer".
sctid:gl002-zsh140412_32_2e4 ctec:hasCUI "C0006142|C0678222".
sctid:gl002-zsh140412_32_2e4 ctec:isA sct:disease_or_syndrome.
sctid:gl002-zsh140412_32_2e5 ctec:isA sct:treatment.
sctid:gl002-zsh140412_32_2e5 ctec:hasTerm "treatment".
sctid:gl002-zsh140412_32_2e5 ctec:hasCUI "C0087111|C1522326|C1533734|C1705169|C3161471".
sctid:gl002-zsh140412_32_2e6 ctec:isA sct:treatment.
sctid:gl002-zsh140412_32_2e6 ctec:hasTerm "reduction".
sctid:gl002-zsh140412_32_2e6 ctec:hasCUI "C0301630|C0392756|C0441610".
sctid:gl002-zsh140412_32_2e5 ctec:hasDrug sctid:gl002-zsh140412_32_2e7.
sctid:gl002-zsh140412_32_2e7 ctec:hasTerm "adjuvant".
sctid:gl002-zsh140412_32_2e7 ctec:hasCUI "C0001551|C0001552|C1522673".
sctid:gl002-zsh140412_32_2e8 ctec:isA sct:EC.
sctid:gl002-zsh140412_32_2e8 ctec:includes sctid:gl002-zsh140412_32_2e3.
sctid:gl002-zsh140412_32_2e8 ctec:hasFragment sctid:gl002-zsh140412_32_2e5.
```

```
sctid:gl002-zsh140412_32_2e8 ctec:hasFragment sctid:gl002-zsh140412_32_2e6.
sctid:gl002-zsh140412_32_2e8 ctec:hasFragment sctid:gl002-zsh140412_32_2e1.
sctid:gl002-zsh140412_32_2e8 ctec:hasFragment sctid:gl002-zsh140412_32_2e2.
sctid:gl002-zsh140412_32_2e8 ctec:hasFragment sctid:gl002-zsh140412_32_2e9.
sctid:gl002-zsh140412_32_2e9 ctec:hasTerm "micrometastases".
sctid:gl002-zsh140412_32_2e9 ctec:hasCUI "C1513276".
sctid:gl002-zsh140412_32_2e9 ctec:hasFragment sctid:gl002-zsh140412_32_2e10.
sctid:gl002-zsh140412_32_2e10 ctec:hasTerm "follow-up".
sctid:gl002-zsh140412_32_2e10 ctec:hasCUI "C0589120|C1522577|C1704685|C3274571".
sctid:gl002-zsh140412_32_2e8 ctec:hasFragment sctid:gl002-zsh140412_32_2e11.
sctid:gl002-zsh140412_32_2 sct:hasGuidelineItemID "gl002-zsh140412_32_2".
```

which states the guideline statement and the relation extractions from the statement. They provide the detailed RDF description of the guideline statement and their annotation with the concepts in UMLS, a well-known meta-thesaurus of medical terms developed [10] and offering mappings to most of widely used medical terminologies such as SNOMED-CT, NCI, LOINC, etc.

3   SemanticCT

SemanticCT[1] is a semantically enabled system for clinical trials. The goals of SemanticCT are not only to achieve inter-operability by semantic integration of heterogeneous data in clinical trials, but also to facilitate automatic reasoning and data processing services for decision support systems in various settings of clinical trials [7]. SemanticCT is built on the top of the LarKC platform[2], a platform for scalable semantic data processing.

The architecture of SemanticCT is shown in Fig. 1. The SemanticCT management component manages the SPARQL endpoint which is built as a SemanticCT workflow which consists of a generic data processing and reasoning plugin in the LarKC platform. That generic data processing and reasoning plug-in provides the basic reasoning service over large scale semantic data, like RDF/RDFS/OWL data. The SemanticCT management component interacts with the SemanticCT Prolog component which provides the rule-based reasoning support in the system.

A screenshot of the interface of SemanticCT is shown in Fig. 2.

SemanticCT provides various semantically-enabled knowledge services for clinical trials. Those knowledge services include (i) Semantic search for SPAQR queries over LarKC SPARQL endpoints; (ii) Automatic patient recruitment by fast identification of eligible patients; (iii) Rule-based reasoning on the formulization of eligibility criteria of clinical trials [6], (iv) Feasibility estimation for a trial design [9], and others. SemanticCT has been designed for the tasks in the European 7th framework project EURECA [3][3].

The guideline update has been considered to be one of the tasks in the EURECA project. The motivation for the guideline update is that clinical guideline update is a time consuming task, which usually take more than six years

---

[1] http://wasp.cs.vu.nl/sct

[2] http://www.larkc.eu

[3] http://eurecaproject.eu

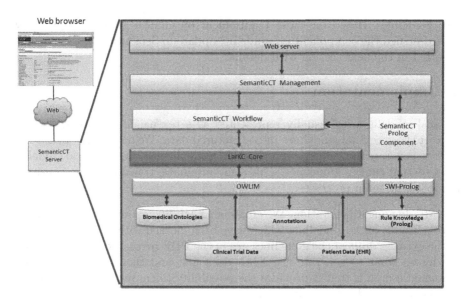

Fig. 1. The architecture of SemanticCT.

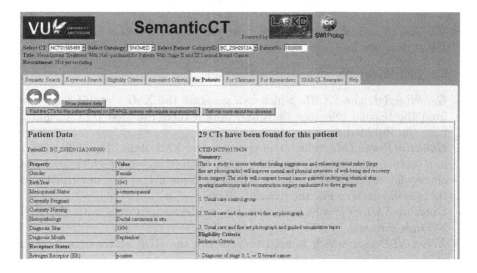

Fig. 2. The interface of SemanticCT

to accommodate new medical findings into guidelines. The task of the guideline update in the EURECA project is to develop the methods so that new evidences can be found more efficiently. That work includes: (i) develop the methods so that existing clinical guidelines can be updated by considering new findings in biomedical researches, (ii) formalize clinical guidelines and align it with the core EURECA dataset enabling thus the semantic interoperability platform to bridge the EHR and clinical trial systems, (iii) transfer new research findings into guidelines. Based on those requirement on clinical guideline update, we have designed the component for guideline update in SemanticCT, which will be reported in the next section with details.

4   Implementation

Clinical guidelines are usually represented as a textual file. Thus, we have to convert those textual guidelines into structured data. We have developed several tools to generate the semantic representation of evidence-based clinical guidelines from textual clinical guidelines. This semi-automatic transformation consists of the following main processes:

- XML document generation. We create an XML document which contains the conclusions of the guidelines which have been marked with the evidence. Since the existing draft of the guidelines are in the PDF textual format. The reason why we want to generate the XML data first and then transfer them into semantic data is that it provides the possibility to convert the guideline data into any other kinds of data formats.
- Evidence statement generation. We use the XSLT tool to convert the XML document into a set of RDF statements which corresponds with the evidence description.
- Guideline statement generation. We use the Xerox NLP tool to generate the RDF statements for each guideline statement.

   The guidelines tools we have developed are:

   - *GuidelineReference2XML*: which can generate the XML data of references from the textual reference data and obtain their PubMed ID (i.e., pmid) from the PubMed website;
   - *GuidelineBody2XML*: which can generate the XML data of guideline items in a textual guideline. We use the textual pattern matching which are based on regular expressions to detect guideline items if they are described with the specific pattern like the example above;
   - *GuidelineXML2NT*: which convert the guideline into ones with the RDF N-Triple format;
   - *Xerox NLP tool*: which is used to generate the RDF statements for each guideline statement.

   We select the Dutch Breast Cancer Guidelines (version 2.0), which has been published in 2012, as the test data [11]. We have converted the complete set of the guideline items in the Dutch Breast Cancer Guidelines into the RDF N-Triple format, and loaded them into the data layer of the LarKC platform and make them integrated with SemanticCT. The basic information of the generated semantic data is shown in Table 1.

   We have implemented a component of evidence-based clinical guidelines in the SemanticCT system. We consider the following main tasks for the update of

Table 1. Generated semantic data of Dutch breast cancer guidelines version 2.0

Item	Number	Triple number
Guideline topic	62	248
Guideline item/recommendation	227	
Evidence	710	4,970
Entity	5,310	
RDF statement		9,744

evidence-based clinical guidelines: (i) guideline browsing, (ii) finding relevant evidence from PubMed for a selected guideline item, (iii) finding relevant evidence from clinical trials for a selected guideline item.

That guideline update interface provides various selection for guideline designer and other researchers to find relevant research findings on guideline content, alternatively called guideline items, or conclusions, or recommendations. The interface provides the selection options to select the formalized evidence-based clinical guidelines with different topics (e.g. subsection title of the guideline document). For each selected topic, the interface will show a list of the guideline items with their evidence levels, refereed literature and their evidence classes.

A screenshot of the interface of the guideline component in SemanticCT is shown in Fig. 3, which shows the topic selection of the component. Figure 4 shows a result of selected topic and their guideline items.

The guideline update component is the tool that helps the guideline developers to search for relevant literature for updates. The tool gives support for formulating the query. The user can select a guideline item, then select different functionalities for checking relevant literature. Those functionalities include:

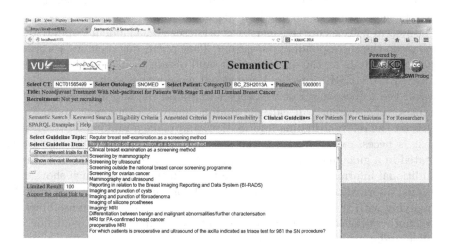

Fig. 3. The guideline component in SemanticCT

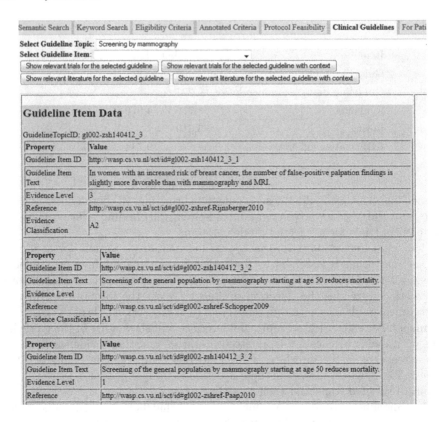

Fig. 4. Guideline items in SemanticCT

find relevant research findings with different options on temporal aspects, such as "latest" new finding or finding in a specific year, and others.

The system will execute the query execution service, a guideline update query service to do the corresponding processing. That processing includes a workflow to make an analysis on the query request, and obtain their annotated concepts such as UMLS/MESH terms, add additional parameters, such as date or year, then post the request to a PubMed server to obtain the information of relevant literatures. The system will call the finding analysis component to make an analysis on the relevant finding information, and make a ranking on the results when it is necessary, then return the selected results to the user. The terms annotated in a guideline item may not be sufficient to find all relevant findings. An additional information is to use those terms in the topics, which show some contextual information of the guideline item. Thus, the component also provides the querying with the context. Those functionalities will show the usefulness of the semantic enriched guideline for browsing and searching in a guideline and guideline update.

5   Use Cases

SemanticCT provides an interface of semantic search, so that the users who have the knowledge of the Semantic Web technology can make SPARQL queries to detect the relations among various data resources in the system. That also provides a basic functionality which is built inside the system to obtain the corresponding results for various requests for ordinary users who may have no any background knowledge of the Semantic Web technology.

Here is a simple example of the SPARQL query which checks the connection (i.e., both use a same concept) between a guideline item and the eligibility criteria of a clinical trial:

```
PREFIX sct: <http://wasp.cs.vu.nl/sct/sct#>
PREFIX ctec: <http://eurecaproject.eu/ctec/>
select distinct ?guidelineid ?guidelinetext ?term ?trialid  ?eligibilityCriteria
where {?s ctec:hasText ?guidelinetext.
?s1 sct:hasRelations ?s.
?s1 sct:hasGuidelineItemID ?guidelineid.
?s ctec:hasFragment ?e1.
?e1 ctec:hasTerm ?term.
?e1 ctec:hasCUI ?conceptid.
?e2 ctec:hasCUI ?conceptid.
?e3 ctec:hasObject ?e2.
?s2 ctec:hasFragment ?e3.
?s2 sct:NCTID ?trialid.
?s2 ctec:hasText ?eligibilityCriteria.}
```

In order to check the relevance between a guideline item and a clinical trial, we usually want to check their connections via multiple concepts. The following is an example of the SPARQL query which checks the connection with at least two concepts:

```
select distinct ?guidelineid ?guidelinetext ?trialid  ?eligibilityCriteria
where {?s ctec:hasText ?guidelinetext.
?s1 sct:hasRelations ?s.
?s1 sct:hasGuidelineItemID ?guidelineid.
?s ctec:hasFragment ?e1.
?e1 ctec:hasTerm ?term1.
?s ctec:hasFragment ?e10.
?e10 ctec:hasTerm ?term2.
FILTER (!(?term1=?term2)).
?e1 ctec:hasCUI ?conceptid1.
?e2 ctec:hasCUI ?conceptid1.
?e3 ctec:hasObject ?e2.
?s2 ctec:hasFragment ?e3.
?e10 ctec:hasCUI ?conceptid2.
?e22 ctec:hasCUI ?conceptid2.
?e32 ctec:hasObject ?e22.
?s2 ctec:hasFragment ?e32.
?s2 sct:NCTID ?trialid.
?s2 ctec:hasText ?eligibilityCriteria.}
```

This gives all trials that share two concepts with the guideline. As defined in Sect. 2, some guideline items of level one are supported by at least two A2 reviews. Thus, we can use the following SPARQL query to find those items:

```
select distinct ?s ?level
where {?s sct:evidenceLevel ?level.
?s sct:hasReferences ?rs.
?rs sct:hasReference ?r1.
?r1 sct:evidenceClassification "A2".
?rs sct:hasReference ?r2.
?r2 sct:evidenceClassification "A2".
FILTER (!(?r1=?r2)).}
```

A similar SPARQL query can be used to check whether or not there exists any item which is supported by at least two A2 reviews, and classified into a non-level-one recommendation, by adding the line '$FILTER(?level > 1)$'. The system can find that there exists one recommendation (gl002-zsh140412_14_3) which are supported by at least two reviews, but classified into one with Level 2. Namely, the following guideline item may contain an error in the guideline:

```
Level 2
The difference in accuracy between MRI and mammography is dependent on the density
of the breast tissue. The difference is small for fatty breasts.
A2 Berg 2004, Sardanelli 2004,
B Van Goethem 2004, Schnall 2005
```

We can also detect some evidences which have marked as a non-standard class. Those errors may result from the semi-automatic processing of the XML file generation. Table 2 shows a summary of checking the relations between recommendations and their evidences. The corresponding figures are shown in Figs. 5 and 6.

We have selected several guideline items in the interface of the guideline component in SemanticCT, and search for their latest relevant finding (i.e., the publication which appears in 2013) for the test. The system can make the analysis

Table 2. Summary of recommendation and evidences

Recommendation	Number	Percentage	Evidence	Number	Percentage
Total recommendation	227	100	Total evidence	710	100
Level 1	94	41.41	A1	125	17.76
Level 1 with at least two A2	45	19.82	A2	171	24.29
Level 2	46	20.26	B	214	30.40
Level 3	87	38.33	C	193	27.41
Level 4	0	0	D	1	0.14
Detected error	1	0.44	Detected error	6	0.85

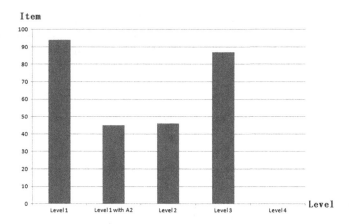

Fig. 5. Recommendation graph

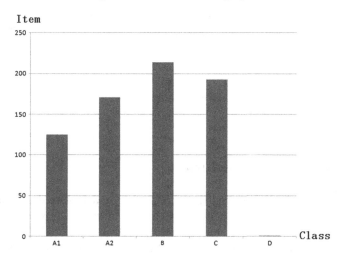

Fig. 6. Evidence graph

of the relations extracted in the guideline text, and create a query to the PubMed Server to obtain their relevant publications. For example, for the guideline item (gl002-zsh140412_2_2) which states that "Clinical breast examination combined with mammography for breast cancer screening has a low sensitivity and a high percentage of false-positive findings, and is therefore not cost-effective.", the system can find a set of UMLS terms (e.g., breast cancer screening, clinical breast examination, sensitivity, and others), then create a request like this:

http://eutils.ncbi.nlm.nih.gov/entrez/eutils/esearch.fcgi?db=pubmed&term =breast cancerscreening[mesh]+AND+Clinicalbreastexamination[mesh]+AND +sensitivity[mesh] +AND+high[mesh]+AND+findings[mesh]+AND+2013[date]

The system can find a set of relevant article (i.e., PMIDs) which appears in 2013 from PubMed: 24860893, 24860794, 24860786, etc. Figure 7 shows a result of relevant evidence (pmid=24860786) for the selected guideline item.

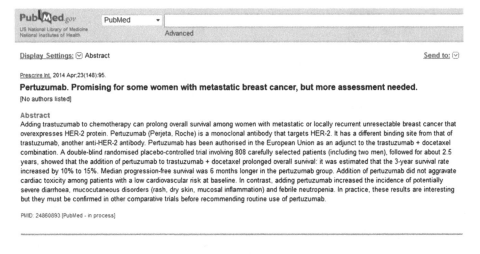

Fig. 7. Relevant evidence in PubMed

6   Discussion and Conclusions

In this paper, we have introduced the semantic approach of evidence-based clinical guidelines in which evidence-based clinical guidelines are represented as the Semantic Web standards such as RDF/RDFS/OWL. We have discussed the design of the guideline component for SemanticCT and shown that how the guideline component has been integrated in SemanticCT. We have described the tools we have developed for the guideline component and how those tools can be used to generate semantic data from textual guidelines into the semantic data. We have reported several use cases on the guideline component and how it is used to find new and relevant findings for selected guideline items. We illustrated that even with this lightweight formalisation we are able to give support to searching/browsing of guidelines, linking guidelines to relevant clinicial trials or to relevant new literature.

What we have done in the experiments is just some initial test with the formalized evidence-based clinical guidelines. Of course, a lot of work is needed to be done to enhance the functionalities of the guideline update and improve the quality of the results. We are going to improve the quality of the relation extraction by the NLP tools, so that it can extract more relations on the guidelines. Another useful issue is that the NLP tools would allow for using our own terminologies so that the annotations can focus on the selected disease, such as breast cancer. In other words, currently we use a rather generic NLP tool, which in the near future we will use a more specialised NLP method because of exploiting medical ontologies and guideline specific structures in the NLP method.

A future experiment is to investigate how the completeness and the soundness of the approach, so that we can improve the proposed approach in this paper. We are also going to develop some methods to do better ranking on the results of the relevant findings. In particular, we are more interested in the logical structure of guideline statements and their logical relations with evidences, so that we can know whether or not a new finding is increasing or decreasing the existing evidence, or is a refutation of the existing evidence. That would improve the efficiency of clinical guideline maintenance significantly.

Acknowledgments. This work is partially supported by the European Commission under the 7th framework programme EURECA Project (FP7-ICT-2011-7, Grant 288048).

References

1. Ait-Mokhtar, S., Bruijn, B.D., Hagege, C., Rupi, P.: Initial prototype for relation identification between concepts. D3.2. Technical report, EURECA Project (2013)
2. Aït-Mokhtar, S., Chanod, J.-P., Roux, C.: Robustness beyond shallowness: incremental deep parsing. Nat. Lang. Eng. **8**(2), 121–144 (2002)
3. Claerhout, B., De Schepper, K., et al.: Initial eureca architecture. D2.2. Technical report, EURECA Project (2013)
4. Fox, J., Johns, N., Lyons, C., Rahmanzadeh, A., Thomson, R., Wilson, P.: Proforma: a general technology for clinical decision support systems. Comput. Methods Programs Biomed. **54**, 59–67 (1997)
5. Fox, J., Johns, N., Rahmanzadeh, A., Thomson, R.: Proforma, approaches for creating computer-interpretable guidelines that facilitate decision support. In: Proceedings of Medical Informatics Europe, Amsterdam (1996)
6. Huang, Z., ten Teije, A., van Harmelen, F.: Rule-based formalization of eligibility criteria for clinical trials. In: Peek, N., Marín Morales, R., Peleg, M. (eds.) AIME 2013. LNCS, vol. 7885, pp. 38–47. Springer, Heidelberg (2013)
7. Huang, Z., ten Teije, A., van Harmelen, F.: SemanticCT: a semantically-enabled system for clinical trials. In: Riaño, D., Lenz, R., Miksch, S., Peleg, M., Reichert, M., ten Teije, A. (eds.) KGC 2013 and ProHealth 2013. LNCS (LNAI), vol. 8268, pp. 11–25. Springer, Heidelberg (2013)
8. Huang, Z., ten Teije, A., van Harmelen, F., Ait-Mokhtar, S.: Semantic representation of evidence-based clinical guidelines. In: Proceedings of 6th International Workshop on Knowledge Representation for Health Care (KR4HC'14) (2014)
9. Huang, Z., van Harmelen, F., ten Teije, A., Dekker, A.: Feasibility estimation for clinical trials. In: Proceedings of the 7th International Conference on Health Informatics (HEALTHINF2014), Angers, Loire Valley, France, 3–6 March 2014
10. Lindberg, D., Humphreys, B., McCray, A.: The unified medical language system. Methods Inf. Med. **32**(4), 281–291 (1993)
11. NABON: Breast cancer, Dutch guideline, version 2.0. Technical report, Integraal kankercentrum Netherland, Nationaal Borstkanker Overleg Nederland (2012)

12. NSRS: Guideline complex regional pain syndrome type i. Technical report, Netherlands Society of Rehabilitation Specialists (2006)
13. Peleg, M., Boxwala, A., Ogunyemi, O., et al.: Glif3: the evolution of a guideline representation format. In: Proceedings of AMIA Annual Fall Symposium, pp. 645–649 (2000)
14. Peleg, M., Boxwala, A., Tu, S., Ogunyemi, O., Zeng, Q., Wang, D.: Guideline interchange format 3.4. Technical report (2001)
15. Shahar, Y., Miksch, S., Johnson, P.: The asgaard project: a task specific framework for the application and critiquing of time oriented clinical guidelines. Artif. Intell. Med. **14**, 29–51 (1998)
16. Tu, S., Musen, M.: A flexible approach to guideline modeling. In: Proceedings of 1999 AMIA Symposium, pp. 420–424 (1999)

Constructing Provenance Cubes Based on Semantic Neuroimaging Data Provenances

Jianhui Chen[1]($\boxtimes$), Jianhua Feng[1], Ning Zhong[2,4], and Zhisheng Huang[3,4]

[1] Department of Computer Science and Technology, Tsinghua University,
Beijing 100084, China
chenjhnh@mail.tsinghua.edu.cn, fengjh@tsinghua.edu.cn
[2] Department of Life Science and Informatics, Maebashi Institute of Technology,
Maebashi-City 371-0816, Japan
zhong@maebashi-it.ac.jp
[3] Department of Computer Science, Vrije University Amsterdam,
Amsterdam, The Netherlands
huang@cs.vu.nl
[4] International WIC Institute, Beijing University of Technology,
Beijing 100124, China

Abstract. The systematic Brain Informatics (BI) study is a data-driven process and all decision-making and suppositions depend on the deep understanding of brain data. Aiming at unstructured brain data, semantic neuroimaging data provenances, called BI provenances, have been constructed to support the quick and comprehensive understanding about data origins and data processing. However, the existing file-based or transaction-database-based provenance queries cannot effectively meet the requirements of understanding data and generating decision or suppositions in the systematic study, which needs multi-aspect and multi-granularity information of provenances. Inspired by the online analytical processing (OLAP) system, this paper proposes provenance cubes to support multi-aspect and multi-granularity provenance queries. A Data-Brain based approach is also designed to develop a BI OLAP system based on provenances cubes. The case study demonstrates significance and usefulness of the proposed approach.

1 Introduction

Brain Informatics (BI) is an emerging interdisciplinary and multidisciplinary research field that focuses on studying the mechanisms underlying the human information processing system [1]. The systematic BI study is a state-of-the-art process. The whole process from experimental designs to result interpretations needs the right decision-making and suppositions from researchers. It is also a data-driven process in which all decision-making and suppositions depend on the deep understanding of brain data, including not only the data obtained from a group of same experiments but also the data obtained from multiple groups of related experiments.

© Springer-Verlag Berlin Heidelberg 2014
D. Zhao et al. (Eds.): CSWS 2014, CCIS 480, pp. 213–226, 2014.
DOI: 10.1007/978-3-662-45495-4_19

Most of brain data are unstructured and cannot be understood directly. Hence, similar to other brain database studies [2], a kind of neuroimaging data provenances, called Brain Informatics provenances [3], has been constructed based on semantic Web technologies, to support the understanding about data origins and data processing. However, understanding data and generating decision or suppositions depend on multi-aspect provenance information of brain data. Researchers also often need to synthetically and repeatedly browse provenance information from different granularities for inspiring new ideas.

Such provenance queries cannot be realized only depending on the existing BI provenances which are stored as files or transactions-oriented databases. This will be complex and time-consuming. In order to complete these provenance queries, researchers have to comprehensively understand scheme and hierarchies of file or databases, and organize query sentences to obtain needed information. Queries also have to be repeatedly performed on massive provenances for each change on information granularities.

Inspired by previous studies on the online analytical processing (OLAP) system, this paper proposes provenance cubes to support multi-aspect and multi-granularity provenance queries. A Data-Brain based approach is also designed to develop a BI OLAP system based on provenances cubes. The remainder of this paper is organized as follows. Section 2 discusses background and related work. Section 3 gives definitions about provenance cubes. Based on the preparations, Sect. 4 proposes a Data-Brain based approach for developing the BI OLAP system. Furthermore, a case study is provided in Sect. 5. Finally, Sect. 6 gives concluding remarks.

2   Background and Related Work

2.1   The Data-Brain and Brain Informatics Provenances

The Data-Brain is a new domain-driven conceptual brain data model, which represents functional relationships among multiple human brain data sources, with respect to all major aspects and capabilities of human information processing system, for systematic investigation and understanding of human intelligence [1]. It includes four dimensions, namely function dimension, data dimension, experiment dimension and analysis dimension, which are connected to each other and corresponding to the four issues of BI methodology, respectively.

Owing to the BI methodology based ontological modeling process, the Data-Brain provides a global conceptual model of systematic BI study. Using the Data-Brain as metadata scheme, BI provenances [3] were proposed to record multi-aspect information about the whole life cycle of brain data, including not only data themselves, data origins (experiments) and data processing (data analysis) but also relationships among experiments and analysis during the systematic BI study. By using semantic Web technologies, the obtained BI provenances are a kind of semantic neuroimaging data provenances.

A BI brain data center [1] has been constructed by integrating the OWL-DL [4] based Data-Brain, RDF [5] based BI provenances and heterogeneous brain data. It is a brain data and knowledge base to integrate valuable data, information and knowledge in the whole BI research process for various data (provenance) requests coming from different aspects of the systematic BI study. By performing the SPARQL [6] query, multi-aspect provenance information of brain data can be obtained to support various decision-making and suppositions.

As stated above, during the data-driven systematic BI study, researchers need to synthetically and repeatedly browse provenance information from different aspects and granularities to generate decision or suppositions. However, such multi-aspect and multi-granularity provenance queries on brain data cannot be effectively supported by the existing BI brain data center. The reasons are as follows.

- Complex query representation: The existing BI brain data center supports multi-aspect and multi-granularity provenance queries by performing SPARQL queries on BI provenances. Because scheme and hierarchies of provenances are based on the ontological Data-Brain, constructing these query sentences needs researchers to understand the structure of Data-Brain in depth. This is difficult for most of researchers.
- Low execution efficiency: Based on the existing BI brain data center, SPARQL queries need to be submitted repeatedly for realizing multi-aspect and multi-granularity provenance queries. Even if a minor change on information granularities, the query sentence still needs to be re-organized and performed on massive BI provenances and the Data-Brain. This is time-consuming and will lead to bad user experiences.

It is necessary to develop a new technology or tool for supporting the multi-aspect and multi-granularity provenance queries on the BI brain data center.

2.2   The Online Analytical Processing System

Comparing with traditional databases, the online analytical processing (OLAP) [7] is an effective approach for processing decision-support and analytical queries by providing three inventions, including object-oriented query representation, virtual cursor and pre-integrated data.

Different from a standard transactional query "When did order 180 ship?", an analytical query might ask, "How do sales in the southeastern region of China for this month comparing with plan or with sales ago?". It is oriented to a specific subject and involves multi-aspect and multi-granularity sale data. Performing this analytical query needs an iterative process. Users begin with a simple query, examine the results, modify the query slightly to highlight an element of interest, and then examine new results of the modified query. This process of incremental modification will be repeated for many cycles. Such an analytical query and its process of incremental modification are the same as multi-aspect and multi-granularity provenance queries on brain data. Hence, developing a BI OLAP system on BI provenances can be an effective approach for supporting multi-aspect and multi-granularity provenance queries which are needed by the systematic BI study.

The development of OLAP includes two main work, data cube modeling, i.e., multidimensional data modeling, and data Extract-Transform-Load (ETL). Both them are state-of-the-art. Some ontology based technologies have been studied to simplify the development of OLAP. For data cube modeling, ontologies can be regarded as a domain conceptual model for automatically identifying potential dimensions and measures based on relations among concepts [8]. For data ETL, ontologies can be regarded as a domain knowledge base for describing data sources and realizing the schemata mapping from data sources to data cubes [9]. The Data-Brain is an ontological model of brain data. Based on it, the BI OLAP system can be developed by an ontology based approach.

3   Provenance Cube

Data cubes are the base of OLAP systems. They are multidimensional data models and used to integrate multi-aspect data about subjects for performing the analytical query. Similarly, the BI OLAP system also includes many data cubes which are used to store multi-aspect provenance information about subjects. In our studies, these data cubes are called "Provenance Cubes" because provenance information is extracted from BI provenances.

This section will give definitions of provenance cubes based on previous study on data cubes [10].

3.1   Basic Definitions

Similar to data cubes, provenance cubes are also subject-oriented. Each subject includes a fact, a group of measures and a group of information dimensions. These components can be defined as follows.

Definition 1. *A Fact, denoted by F, is a decision subject or a research focus, such as the brain activation, which is interested by investigators during the systematic BI study.*

Definition 2. *A Measure, denoted by M, is an attribute of F, which describes a characteristic of the F itself, such as the peak of brain activations.*

Definition 3. *An Information Dimension, denoted by ID, is a specific description aspect of F, such as the cognitive function corresponding to the brain activation, and has a tree structure including many levels. Each level is a Dimension Level, denoted by DL, which includes some nodes and represents a specific abstract level or granularity of ID. Each node is a Dimension Member, denoted by DM, which represents a specific value on the corresponding abstract level of ID. The set of dimension members in the information dimension ID_i is denoted by $membs(ID_i)$, in which the set of dimension members in the k-th dimension level is denoted by $membs(ID_i)(k)$ (when $k = 1$ is the lowest dimension level of ID_i).*

Definition 4. *An Information Dimensional Space, denoted by $IDS = \{ID_1, ID_2, ..., ID_n\}$, is the set of all possible information dimensions in the systematic BI study.*

3.2   The Definition of Provenance Cubes

In order to store those fact instances, a provenance cube should include all components of subjects, including the fact, measures and information dimensions. Based on the above basic definitions, it can be defined as follows.

Definition 5. *A Basic Provenance Cube, denoted by $PCube_b$, is a three-tuple:*

$$(ID_b, M_b, R_b),$$

where,

- $ID_b = (ID_1, ID_2, ..., ID_n)$ *is an information dimension list in which each $ID_i (i = 1, 2, ..., n)$ is an information dimension;*
- $M_b = (M_1, M_2, ..., M_v)$ *is a measure list in which each $M_k (k = 1, 2, ..., v)$ is a measure;*
- $R_b = \{cell_1, cell_2, ..., cell_t\}$ *is a set of information cells in which each $cell_j (j = 1, 2, ..., t) = (x_{j1}, x_{j2}, ..., x_{jn}, m_j)$ is used to store a fact instance, where $x_{ji} = cell_j[ID_i] = $ an instance of $DM_l \wedge DM_l \in membs(ID_i)(1)$ is the value of information dimension ID_i in $cell_j$ and $m_j = (m_{j1}, m_{j2}, ..., m_{jv})$ is the measure value list of $cell_j$ in which each m_{jk} is the value of measure M_k in $cell_j$.*

Definition 6. *A Provenance Cube, denoted by $PCube$, is a six-tuple:*

$$(ID, DL, M, PCube_b, A, R),$$

where,

- $ID = (ID_1, ID_2, ..., ID_n)$ *is an information dimension list;*
- $DL = (DL_1, DL_2, ..., DL_n)$ *is a dimension level list in which each $DL_i (i = 1, 2, ..., n)$ is the dimension level of information dimension ID_i;*
- $M = (M_1, M_2, ..., M_v)$ *is a measure list;*
- $PCube_b$ *is the basic provenance cube which is used to create $PCube$;*
- $A = (A_1, A_2, ..., A_v)$ *is an aggregation function list in which each A_k is the aggregation function of measure M_k;*
- $R = \{cell_1, cell_2, ..., cell_t\}$ *is a set of information cells in which each $cell_j (j = 1, 2, ..., t) = (x_{j1}, x_{j2}, ..., x_{jn}, m_j)$ is used to store an aggregation result of fact instances, where $x_{ji} = cell_j[ID_i] \in membs(ID_i)(DL_i)$ is the value of information dimension ID_i in $cell_j$ and $m_j = (m_{j1}, m_{j2}, ..., m_{jv})$ is the measure value list of $cell_j$ in which each m_{jk} is the value of measure M_k in $cell_j$.*

A basic provenance cube is shown in Fig. 1(a). It can be denoted by $PCube_b = (ID_b, M_b, R_b)$, where $ID_b = \{ID_{Cognitive-Function}, ID_{Brain-Region}, ID_{Experimental-Task}\}$ indicates that $PCube_b$ includes three information dimensions *Cognitive-Function*, *Brain-Region* and *Experimental-Task*, corresponding to the three coordinate axes; $M_b = (Nox, Peak)$ indicates that the fact *Brain-Activation* includes two measures, namely, *Nox* (the size of activation) and *Peak* (the coordinate of activation center), for describing characteristics

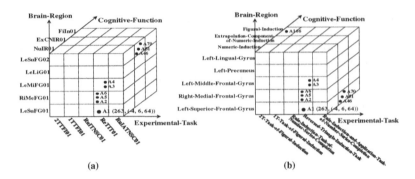

Fig. 1. The basic provenance cube and provenance cube for the subject "About brain activations": (a) A basic provenance cube (b) A provenance cube.

of brain activations themselves; $R_b = \{A1, A2, ..., A166\}$ indicates that $PCube_b$ stores 166 fact instances though many activations cannot be shown because of the limitation of spaces, where $A1 = $ (http://www.wici.org/BIProvenances/Numeric-Induction/NuIR01, http://www.wici.org/BIProvenances/Left-Superior-Frontal-Gyrus/LeSuFG01, http://www.wici.org/BIProvenances/Reversed-Triangle-Induction-Task/ReTIT01, $(263, (-4, 6, 64))$). As shown in this figure, all scale values on three coordinate axes are instances of concepts of the Data-Brain, which can be directly extracted from BI provenances. For example, $ReTIT01$ on the coordinate axis $Experimental\text{-}Task$ is an instance of the experimental task concept $Reversed\text{-}Triangle\text{-}Induction\text{-}Task$ and represents the experimental task stated in our previous study [11].

Figure 1(b) is a provenance cube which was created by performing once aggregation operation on three information dimensions of $PCube_b$. It can be denoted by $PCube = (ID_b, DL, M_b, PCube_b, A, R)$, where $DL = (1, 1, 1)$ indicates that all dimension levels are 1; $A = (push(), push())$ indicates that the aggregation functions of both measures Nox and $Peak$ are the operation $push()$ which pushes measure values into the corresponding Nox or $Peak$ stack; $R = \{A1, A2+A5+A6, ..., A166\}$ indicates that $PCube$ stores 166 fact instances, where $A1 = $ (http://www.wici.org/Data-Brain/FunctionDimension.owl#Numeric-Induction, http://www.wici.org/Data-Brain/AnalysisDimension.owl#Left-Superior Frontal-Gyrus, http://www.wici.org/Data-Brain/ExperimentDim-ension.owl# Reversed-Triangle-Induction-Task, $(263, (-4, 6, 64))$). As shown in this figure, all scale values on three coordinate axes are the lowest level of concepts in the Data-Brain because of $DL = (1, 1, 1)$.

3.3 A Comparison Between Provenance Cubes and Traditional Data Cubes

Provenance cubes are a kind of special data cubes which are used to store BI provenance information for supporting the multi-aspect and multi-granularity provenance queries on brain data. Comparing the above definitions of provenance cubes with traditional data cubes, the most important difference is on information dimensions.

Traditional data cubes define the dimension as a "Lattice". As shown in Fig. 2(a), each dimension lattice includes a lot of nodes called dimension attributes, such as "Year", "Quarter" and "Month". For each fact instance stored in the data cube, its dimension values are values of dimension attributes, e.g. "2008 year". Different from the dimension of traditional data cubes, the information dimension of provenance cubes is a tree structure shown in Fig. 2(b). Each information dimension includes a lot of nodes called dimension members, such as "Problem-Solving", "Reasoning" and "Deduction". For each fact instance stored in the provenance cube, values of information dimensions are just those dimension members rather than values of dimension members.

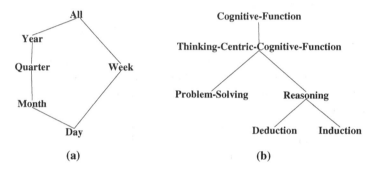

Fig. 2. The dimension and information dimension: (a) A dimension of data cubes (b) An information dimension of provenance cubes.

The reason for this change is as follows. Provenance cubes are used to store provenance information extracted from BI provenances. Because BI provenances consist of instances of concepts of the Data-Brain, these instances include the smallest granularity of provenance information and different levels of concepts in the Data-Brain represent the higher granularities of provenance information. Furthermore, as stated in our previous study [1], the Data-Brain includes four dimensions to model the four core issues of systematic BI methodology, respectively. Each dimension includes a or several sub-dimensions which consist of different levels of concepts and are used to model some specific aspects of the systematic BI study, e.g., cognitive functions. Hierarchies of sub-dimensions represent changes of research viewpoints on different granularities. Hence, the hierarchies of sub-dimensions in the Data-Brain can be used as information dimensions of provenance cubes. Defining the information dimension as a tree structure, just like the conceptual hierarchy, makes it possible to directly map hierarchies of sub-dimensions in the Data-Brain to information dimensions of provenance cubes. A Data-Brain based approach for developing the BI OLAP system can be realized.

According to this change on information dimensions, the above definitions of provenance cubes have been revised based on the definitions of traditional data cubes.

4   A Data-Brain Based Approach for Developing the BI OLAP System

Based on the above definitions of provenance cubes, the BI OLAP system can be developed by a Data-Brain based approach. It includes three steps, namely, Data-Brain based dimensional space construction, Data-Brain based provenance cube modeling, and Data-Brain driven data ETL.

4.1   Data-Brain Based Dimensional Space Construction

The Data-Brain based dimensional space construction is the first step for developing a BI OLAP system. As stated above, hierarchies of sub-dimensions in the Data-Brain can be used as information dimensions of provenance cubes. Hence, this step is just to extract these hierarchies from the Data-Brain based on the hierarchical relations *is-a* and *part-of*. Each sub-dimension is corresponding to an information dimension in which each dimensional member is corresponding to a concept in the sub-dimension. All sub-dimensions in the Data-Brain form the whole information dimensional space for developing the BI OLAP system.

The dimensional space construction is independent of subjects. The BI OLAP system only includes an information dimensional space which can support various provenance cube modeling for different subjects.

4.2   Data-Brain Based Provenance Cube Modeling

The Data-Brain based provenance cube modeling is the second step for developing a BI OLAP system. Based on the aforementioned definition of provenance cube, provenance cube modeling needs to identify information dimensions and measures. In previous studies of ontology based data cube modeling [8], dimensions and measures were identified automatically based on non-hierarchical relations among concepts, i.e., all relations except for *is-a* and *part-of*. Different from these studies, provenance cube modeling is oriented to the single data source, namely, BI provenances, and all of potential decision subjects or research focuses have been defined in the Data-Brain because of the BI methodology based modeling approach. Thus, the Data-Brain based provenance cube modeling is a brand-new approach for ontology based data cube modeling and includes the following four sub-steps:

- identifying a fact concept F from the Data-Brain based on the subject, and transforming its data type properties as measures to define a measure list M_b,
- choosing the corresponding information dimensions from the information dimension space based on requirements of provenance queries to define an information dimension list ID_b,

- identifying the initial dimension level of each information dimension and the aggregation function of each measure based on requirements of provenance queries to define DL and A,
- getting an initial provenance cube model $PCube_m = (ID_b, DL, M_b, (ID_b, M_b, \emptyset), A, \emptyset)$.

These steps can be automatic. The identifying of F and DL can be realized by matching some key concepts which are extracted from the description of requirements of provenance queries. The identifying of ID_b can be realized by searching concepts based on non-hierarchical relations in the Data-Brain. However, the obtained provenance cube model will be intricate and include many undesired information dimensions and measures by such an automatic approach. Hence, this study adopts a manual approach to complete the above steps.

4.3   Data-Brain Driven Data ETL

The Data-Brain driven data ETL is the third step for developing a BI OLAP system. It is just to extract provenance information from BI provenances and store it into provenance cubes. Previous studies on the ontology based data ETL [9] mainly focused on data integration and schemata mapping. Different from previous studies, the data ETL for the BI OLAP system is oriented to BI provenances which have the uniform format, and focuses on automatic information extraction. The whole process is Data-Brain-Driven and includes the following four sub-steps:

- Searching paths: It is to define a dimension concept set DC which includes the top member of each information dimension in the information dimension list ID_b, and then find traversal pathes from the fact concept F to each dimension concept in DC based on the Data-Brain. In the weighted directed graph which are created by using concepts and relationships among concepts in the Data-Brain, those obtained traversal pathes are just the shortest simple pathes between F and each concept in DC. All obtained pathes will be stored into a path set *routes*.
- Realizing the initial provenance cube model $PCube_m = (ID_b, DL, M_b, (ID_b, M_b, \emptyset), A, \emptyset)$: It is to realize $PCube_m$ on relational databases by using the star schema [12]. This can be an automatic process and all database tables can be created based on the fact concept and dimension concepts in the Data-Brain.
- Extracting provenance information: It is to define a SPARQL query for extracting the needed provenance information from BI provenances. The SPARQL query can be automatically created. Its "SELECT" clause can be defined based on the fact concept F, the dimension concept set DC and the measures list M_b, and its "WHERE" clause can be defined based on the measure list M_b and the path set *routes*.
- Realizing the provenance cube $PCube$: It is to fill the start schema of $PCube_m$ by using obtained query results. Because provenance information is some instances of concepts of the Data-Brain, directly filling $PCube_m$ using query

results will obtain not $PCube = (ID_b, DL, M_b, (ID_b, M_b, R_b), A, R \neq \emptyset)$ but $PCube' = (ID_b, DL', M_b, (ID_b, M_b, R_b), A, R \neq \emptyset), where DL' = (0, 0, 0, ...)^1$. Hence, the operation $Roll\text{-}up(PCube', ID_b, DL)$ also needs to be performed to realize the provenance cube $PCube$, i.e., $PCube = Roll\text{-}up(PCube', ID_b, DL)$.

5   A Case Study

In this section, a provenance cube will be constructed to support the interpretation of the activation $A1(Nox(263), Peak(-4, 6, 64))$, which was found in the reversed triangle induction task [11]. This realistic use case is used to illustrate the proposed provenance cubes and the Data-Brain based developing approach of BI OLAP system. Experiments are based on a prototype system of brain data center stated in our previous study [1], which includes human inductive reasoning centric BI experimental data, an induction centric Data-Brain prototype, and the corresponding BI provenances.

In order to interpret the activation $A1$, it is necessary to browse other brain activations found in induction-centric systematic studies. The needed information should be multi-aspect, involved with not only brain activations themselves (e.g. peak, size) but also related cognitive functions, brain regions and experimental tasks. The needed information should also be multi-granularity, involved with different levels of cognitive functions, brain regions and experimental tasks. For supporting such multi-aspect and multi-granularity provenance queries, a provenance cube can be constructed by the above Data-Brain based approach.

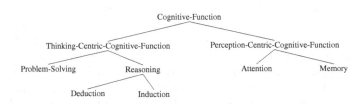

Fig. 3. The information dimension $ID_{Cognitive-Function}$.

The first step is the Data-Brain based dimensional space construction. Each sub-dimension of the Data-Brain can be used to create an information dimension. For example, based on a sub-dimension in the Data-Brain, the information dimension $ID_{Cognitive-Function}$ can be obtained. As shown in Fig. 3, it includes four dimension levels and $membs(ID_{Cognitive-Function})(1) = \{Deduction, Induction\}$. Similarly, $ID_{Brain-Region}$ and $ID_{Experimental-Task}$ can also be obtained.

---

1 As stated above, "1" means the lowest dimension level of information dimensions. In this Data-Brain driven data ETL, information dimensions are corresponding to sub-dimensions of the Data-Brain. Thus, the extracted provenance information is instances of the lowest level of dimensional members and has the corresponding dimension level "0".

The second step is the Data-Brain based provenance cube modeling. A provenance cube model $PCube_m = (ID_b, DL, M_b, (ID_b, M_b, \emptyset), A, \emptyset)$ can be obtained, where $ID_b = (ID_{Cognitive-Function}, ID_{Brain-Region}, ID_{Experimental-Task})$, $DL = (1, 1, 1)$, $M_b = (Nox, Peak)$ and $A = (push(), push())$. $DL = (1, 1, 1)$ means that the lowest level is set as the initial dimension level of each information dimension for storing brain-activation-related provenance information with the smallest granularity.

The third step is the Data-Brain driven data ETL. A dimension concept set $DC = \{Cognitive-Function, Brain-Region, Experimental-Task\}$ can be defined and a path set $routes = \{route_{Brain-Activation->Cognitive-Function},$ $route_{Brain-Activation->Brain-Region}, route_{Brain-Activation->Experimental-Task}\}$ can be gotten by searching paths on the induction centric Data-Brain prototype [1], where $route_{Brain-Activation->Cognitive-Function} = ((Brain-Activation,$ $produced-by, Experimental-Task), (Experimental-Task, has-experimental-purpose, Cognitive-Function))$ is the traversal path from the fact concept $Brain-Activation$ to the dimension concept $Cognitive-Function$. Based on the fact concept and the three dimension concept, the initial provenance cube model $PCube_m$ can be realized on relational databases by using the star schema shown in Fig. 4. A SPARQL query shown in Fig. 5 can also be automatically defined based on F, DC, M_b and $routes$, to extract provenances information about 166 brain activations, involved with reversed triangle tasks [13], number series induction tasks [14] and figural tasks [15]. Finally, $PCube' = (ID_b, DL', M_b,$ $(ID_b, M_b, R_b), A, R)$, where $R = \{A1, A2, ..., A166\}$, can be obtained by using provenance information to fill the start schema of $PCube_m$, and the provenance cube $PCube$ can also be realized by perform the operation $Roll-up$ on $PCube'$. Partial activations in $PCube'$ and $PCube$ are just shown in Fig. 1(a) and (b). As shown in this figure, some activations, such as $A2, A5$ and $A6$, are included in the same information cell of $PCube$ because they have the same dimension value on

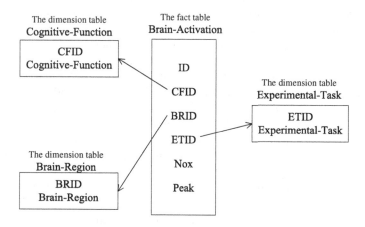

Fig. 4. The star schema for realizing $PCube_m$.

```
Q1:SELECT DISTINCT ?Brain-Activation ?Nox ?Peak ?Cognitive -Function
?Brain-Region ?Experimental-Task WHERE {
?Brain-Activation db:Nox ? Nox.
?Brain-Activation db:Peak ?Peak.
?Brain-Activation rdf:type db:Brain-Activation.
?Brain-Activation db:produced-by ?Experimental-Task.
?Experimental-Task rdf:type db:Experimental-Task.
?Experimental-Task db:has-experimental-purpose ?Cognitive-Function.
?Cognitive-Function rdf:type db:Cognitive-Function.
?Brain-Activation db:located-in ?Brain-Region.
?Brain-Region rdf:type db:Brain-Region
} ORDER BY ASC(?Brain-Activation)
```

Fig. 5. A SPARQL query for extracting provenance information.

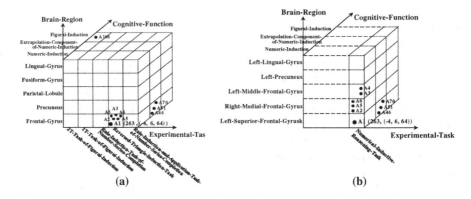

Fig. 6. The partial activations of $PCube_1$ and $PCube_2$: (a) $PCube_1$ (b) $PCube_2$.

each information dimension. Hence, $PCube$ only includes 86 information cells, i.e., its fact table only includes 86 records.

Based on $PCube$, different OLAP operations, including *roll-up*, *drill-down*, *dice* and *slice*, can be performed to support multi-aspect and multi-granularity provenance queries during the interpretation of the activation $A1$.

For finding the similarity and difference of experimental tasks among nearby brain activations, researchers should focus on the aspect "brain regions" and upgrade its granularity from the "Left Superior frontal gyrus" level to the "Frontal gyrus" level. The needed provenance information can be obtained by performing the operation $Roll\text{-}up(PCube, id, dl') = Roll\text{-}up(PCube, (ID_{Brain-Region}), (2))$ on $PCube$ to create a new cube $PCube_1$ shown in Fig. 6(a). $PCube_1$ only includes 53 information cells which store 166 activations. Many activations, such as $A1$, $A2$, $A3$, $A4$, $A5$ and $A6$, have to be included in the same information cell because of the higher granularity on brain regions.

For summarizing the similarity and difference of brain activations among different numerical inductive reasoning tasks, researchers should focus on another aspect "experimental tasks" and its a special case "numerical inductive reasoning task". The needed provenance information can be obtained by performing the composition operation $Dice(Roll -up(PCube, id, dl'), id, \sigma(id, v)) = Dice$

($Roll\text{-}up$ ($PCube$, (id), (2)), (id), $(= (id, Numerical\text{-}Inductive\text{-}Reasoning\text{-}Task)))$ on $PCube$, where $id = (ID_{Experimental-Task})$, to create a new cube $PCube_2$ shown in Fig. 6(b). $PCube_2$ only includes 41 information cells which store 68 activations. 98 activations were deleted because of the operation *dice* on the information dimension $ID_{Experimental-Task}$.

6   Conclusions

This paper proposed provenance cubes and a Data-Brain based approach to develop a BI OLAP system, which can effectively support multi-aspect and multi-granularity provenance queries for the systematic BI study. Comparing with previous studies on the ontology based OLAP system development, the proposed approach has the following advantages.

- Optimizing multi-dimensional modeling: Previous related studies identified facts and dimensions using non-hierarchical relations because they are based on the ontologies coming from database schemata or business models. In these ontologies, major relations are non-hierarchical relations. Different previous studies, provenance cube modeling is based on the ontological Data-Brain in which major relations are hierarchical relations. Thus, our study redefines the information dimension and identifies information dimensions using hierarchical relations. Such a change simplifies the modeling process and improves the quality of obtained dimensions. In fact, most of domain ontologies in BI and biomedical domain, such as gene ontology, are term ontologies and have the similar structure with the Data-Brain. Hence, the proposed Data-Brain based approach is more fit for ontology based multi-dimensional modeling in BI and biomedical domain.
- Simplifying data ETL: Previous studies on ontology based data ETL mainly focused on data integration and schema mapping because they were oriented to heterogeneous data sources. Different from previous studies, the development of BI OLAP system is oriented to RDF based BI provenances which have an uniform data format. Thus, the Data-Brain driven data ETL focuses on data extraction and storage. Both the SPARQL sentence and the star schema can be created automatically based on the Data-Brain. This simplifies the data ETL for developing a BI OLAP system.

Since this project is very new, only some preliminary results were obtained. The future work will focus on identifying latent information dimensions based on non-hierarchical relations among concepts for enriching analytical viewpoints on BI provenances.

Acknowledgments. The work is supported by National Basic Research Program of China (2014CB744600), China Postdoctoral Science Foundation (2013M540096), International Science & Technology Cooperation Program of China (2013DFA32180), National Natural Science Foundation of China (61272345), Open Foundation of Key Laboratory of Multimedia and Intelligent Software (Beijing University of Technology), Beijing.

References

1. Zhong, N., Chen, J.H.: Constructing a new-style conceptual model of brain data for systematic brain informatics. IEEE Trans. Knowl. Data Eng. **24**(12), 2127–2142 (2011)
2. MacKenzie-Graham, A.J., Horn, J.D.V., Woods, R.P., Crawford, K.L., Toga, A.W.: Provenance in neuroimaging. NeuroImage **42**(1), 178–195 (2008)
3. Chen, J.H., Zhong, N., Liang, P.P.: Data-brain driven systematic human brain data analysis: a case study in numerical inductive reasoning centric investigation. Cogn. Syst. Res. Int. J. **15**(16), 17–32 (2012)
4. McGuinness, D.L., Harmelen, F.V.: Owl web ontology language overview. Technical report, World Wide Web Consortium (W3C) recommendation (2004). http://www.w3.org/TR/owl-features/
5. Klyne, G., Carroll, J.J.: Resource description framework (rdf): concepts and abstract syntax. Technical report, World Wide Web Consortium (W3C) recommendation (2004). http://www.w3.org/TR/rdf-concepts/
6. Prud'hommeaux, E., Seaborne, A.: Sparql query language for rdf. Technical report, World Wide Web Consortium (W3C) recommendation (2008). http://www.w3.org/TR/rdf-sparql-query/
7. Greenfield, D., Lyon, G.F., Vogl, R., Feinstein, S.: System and method for online analytical processing. Technical report 7,010,523, Google Patents (2006)
8. Romero, O., Abello, A.: A framework for multidimensional design of data warehouses from ontologies. Data Knowl. Eng. **69**(11), 1138–1157 (2010)
9. Skoutas, D., Simitsis, A.: Ontology-based conceptual design of etl processes for both structured and semi-structured data. Int. J. Seman. Web Inf. Syst. **3**(4), 1–24 (2007)
10. Vassiliadis, P.: Modeling multidimensional databases, cubes and cube operations. In: Proceedings of 10th International Conference on Scientific and Statistical Database Management (SSDBM), Capri, Italy, pp. 53–62 (1998)
11. Liang, P., Zhong, N., Lu, S., Liu, J., Yao, Y., Li, K., Yang, Y.: The neural mechanism of human numerical inductive reasoning process: a combined ERP and fMRI study. In: Zhong, N., Liu, J., Yao, Y., Wu, J., Lu, S., Li, K. (eds.) WImBI 2006. LNCS (LNAI), vol. 4845, pp. 223–243. Springer, Heidelberg (2007)
12. Chaudhuri, S., Dayal, U.: An overview of data warehousing and olap technology. ACM SIGMOD Rec. **26**(1), 65–74 (1997)
13. Lu, S.F., Liang, P.P., Yang, Y.H., Li, K.C.: Recruitment of the pre-motor area in human inductive reasoning: an fmri study. Cogn. Syst. Res. Int. J. **1**(1), 74–80 (2010)
14. Jia, X.Q., Liang, P.P., Lu, J., Yang, Y.H., Zhong, N., Li, K.C.: Common and dissociable neural correlates associated with component processes of inductive reasoning. NeuroImage **56**(4), 2292–2299 (2011)
15. Mei, Y., Liang, P.P., Lu, S.F., Zhong, N., Li, K.C., Yang, Y.H.: Neural mechanism of figural inductive reasoning: an fmri study. Acta Psychol. Sin. **42**(4), 496–506 (2010)

Monitoring Urban Waterlogging Disaster Using Social Sensors

Ningyu Zhang, Guozhou Zheng$^{(\boxtimes)}$, Huajun Chen, Xi Chen, and Jiaoyan Chen

Department of Computer Science, Zhejiang University, Hangzhou, China
{zhangningyu,zzzgz}@zju.edu.cn

Abstract. Nowadays, urban waterlogging has been one of the most serious global urban hazards in some big cities in the world especially in Chinese cities. While, existing methods fail to cover all locations and forecast the waterlogging trend. Meanwhile, the past one decade has witnessed an astounding outburst in the number of online social media services. For example, when a rainstorm occurs, people make a large number of tweets related to the rainstorm, which enables detection of urban waterlogging promptly, simply by analyzing the tweets. In this paper, we present a semantic method that can monitor urban waterlogging using social sensors. Currently, we use ontology and fuzzy reasoning to analyze waterlogging locations and its severity and build Apps to monitor and forecast waterlogging in more than ten cities in China. With this method, people can easily monitor all the possible urban waterlogging locations with severity and trend, which may reduce the possibility of traffic congestion in a rainstorm.

1 Introduction

Urban waterlogging has been one of the most serious hazards in China in these years. In 2013, according to statistics, urban waterlogging attacked in over 100 cities including Beijing, Shanghai, Guangzhou in China. Especially the latest and largest urban waterlogging in Beijing was the so-called "7.21" catastrophic natural disaster, which leads to such consequences as dozens of casualties [13]. There are already several limited ways to monitor the urban waterlogging such as automated water level meter and traffic cameras. However, existing methods fail to cover all locations and forecast the waterlogging trend.

Recently, researchers consider the twitter posts (i.e., tweets) as real-time social data and focus on analyzing the features of keywords in the specific context [12,14,19,22,24]. In addition, there are a number of efforts to exploit the spatial information to detect events, i.e., public events or emergency situations. For example, an earthquake reporting system using Japanese tweets [10,17], EvenTweet: a system to detect localized events from a stream of tweets in real-time [1], Backstrom: a probabilistic framework for quantifying spatial variation [4] and so on [9,16,18,21,23,25].

© Springer-Verlag Berlin Heidelberg 2014
D. Zhao et al. (Eds.): CSWS 2014, CCIS 480, pp. 227–236, 2014.
DOI: 10.1007/978-3-662-45495-4_20

Fig. 1. Android application

Actually, a huge amount of content is generated and shared by social sensors every day [2,5,11,20,26]. In Weibo[1], a twitter-like social site in China, most of the user's profile include the user's self-reported geographic location [8,15]. For example, people who post tweets by mobile phones in Weibo can share their real locations. When a rainstorm occurs, people make a large number of tweets related to it. Huge social data provide a data source forming a large number of social sensors [3,6,7].

In this paper, we present a system that can monitor urban waterlogging using tweets. Our system can analyze the real or history waterlogging condition. We can show the pictures of waterlogging locations, assess their severity and make the prediction of waterlogging trend. We have already developed a web site and an App[2] to demonstrate the system as showed in Fig. 1. Now, our system can track the urban waterlogging in more than ten big cities in China. We use ontology to make a semantic analysis of tweets and retrieve all the waterlogging locations. Then, we use fuzzy reasoning to assess the severity of waterlogging. At last, by analyzing historical data and real tweets stream we forecast the trend.

---

[1] http://weibo.com/

[2] Scan the two-dimensional code to download Apps http://www.cheosgrid.cn:8000/.

English Version   Chinese Version

When a rain-storm occurs, we can provide users with real-time street conditions and send warning message. In future, we will try to analyze twitter data to track more cities in the whole world.

2   Approaches

2.1   Overview

Modern social data analysis for urban waterlogging faces a confluence of growing challenges and problems:

(1) First, we need to detect the locations of waterlogging site accurately based on a large amount of tweets. At present, there is a great deal of method to detect locations of events. However, most are based on the keyword method, which cannot accurately obtain semantic information.
(2) Second, we need to evaluate the severity of waterlogging based on context analysis of tweets, which is helpful for traffic grooming. However, there is no method to judge the waterlogging severity at present.
(3) Finally, we need to predict trends based on historical data. Trend forecast of events can be solved by the statistics and machine learning method.

To tackle these problems, in this work, we present a system that can monitor urban waterlogging using social sensors.

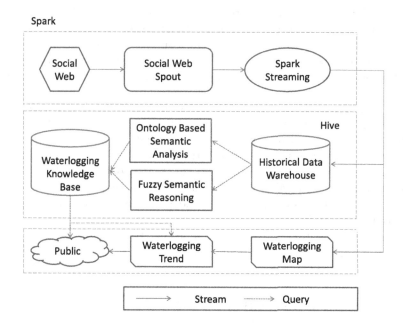

Fig. 2. System architecture

We detail our strategy by utilizing the flow chart is shown in Fig. 2. To get a better understanding of how our system works, the complete system can be broken down into three modules:

(1) Ontology based event analysis: Tweets are collected continuously using the Sina API. We convert tweets to RDF whose schemas is designed as a metadata data model, and store them into the history database. In addition, the system can filter and store relevant information about urban waterlogging;
(2) Fuzzy severity evaluation: at different locations have different severity of waterlogging we need to evaluate these tweets using fuzzy semantic method;
(3) Real-time trend predation: Because twitter posts (i.e., tweets) are real-time social streams, we use Spark streaming to analyze those stream data. The prediction results are pushed to the Apps.

2.2   Ontology Based Event Analysis

Microblog is a complex combination of statements. Ontology can be used in an integration task to describe the semantics of the information sources and to make the content explicit. Simple Knowledge Organization System (SKOS) provides a data model and vocabulary for expressing Knowledge Organization Systems (KOSs) such as thesauri and classification schemes in Semantic Web applications. To filter out the information containing waterlogging locations and time information, we use SKOS to retrieve waterlogging tweets of huge social data.

Once collected, they are converted into RDF triples providing a description of events using the LODE ontology and a large SKOS taxonomy of event categories. LODE is a minimal model that encapsulates the most useful properties for describing events. The goal of this ontology is to enable interpretable modeling of the factual aspects of events, where these can be characterized in terms of the four Ws: What happened, where did it happen, when did it happen, and who was involved. The dataset contains a highly diverse set of urban waterlogging categories. It also provides the description of media using the W3C Ontology for Media Resources and uses properties from SIOC, FOAF, Dublin Core and vCard. Figure 3 depicts the metadata attached to the waterlogging identified by 3163952 on Sina Weibo according to the LODE ontology. More precisely, it indicates that waterlogging has been occurred on the 22th in July 2013 at 18:33 in the Tiananmen Square. This waterlogging is matched with a similar one announced on upcoming.

To conclude, of ontology based event analysis, we get the waterlogging accurate information from large amount of tweets. We store these data into the waterlogging knowledge base.

2.3   Fuzzy Severity Evaluation

There are many words to describe the severity of waterlogging, so making the assessment of waterlogging severity precisely is difficult. We use fuzzy sets to

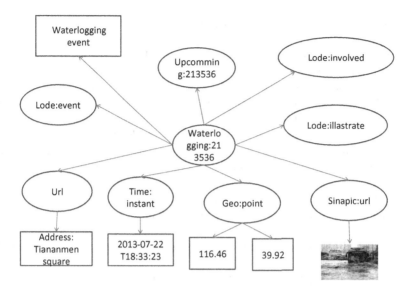

Fig. 3. LODE ontology

describe these words. The fuzzy sets theory provides a framework for the representation of the uncertainty of many aspects of human knowledge. For a given element, fuzzy sets theory proposes the use of intermediate degrees of membership in a set. In this way, if we consider the set of urban waterlogging events we can consider that a location which is severe belongs to such a set with a degree of 1 (belongs), a location which is not very severe belongs in some degree (for example 0.7) and a location which are dry do not belong to this set (degree of 0).

We use the steps below to reason the severity of urban waterlogging locations. First, obtain the location description characteristics established based on LOAD ontology. Then, the values are converted into the fuzzy sets to describe the waterlogging characteristics. The fuzzy value for each variable is obtained by means of the membership function related to each fuzzy set. Once the waterlogging description has been defuzed, the fuzzy rules are evaluated obtaining the set of fuzzy values. The results obtained are defuzed in order to obtain a set of concrete characteristics for the locations. With the defuzed results, locations with the obtained characteristics are retrieved based on the LOAD ontology and the annotations. Next the fuzzy system will calculate the weight of each location based on the values of the suitability obtained from the decision matrix.

Totally, we can evaluate the severity of waterlogging based on fuzzy analysis of tweets and store those data into waterlogging knowledge base.

2.4   Real-Time Trend Prediction

Twitter posts are usually real-time social streams. Spark Streaming is an extension of the core Spark API that allows to enable high-throughput, fault-tolerant stream processing of live data streams. We use Spark Streaming to retrieve the locations of urban waterlogging in a time span and forecast the trend.

To get the waterlogging locations is quite easy from waterlogging knowledge base. To forecast the trend of waterlogging locations, we classify a tweet into a positive class or a negative class. According to the current keywords in a tweet and waterlogging condition in history, we make the classification with statics. We use the steps below to measure trends. By preparing positive and negative examples as a training set, we can produce a model to classify tweets from the same location automatically into positive and negative categories. In positive categories, the waterlogging is gradually weakened. In negative categories, the waterlogging is gradually strengthened. By training the sample data, the system will predict whether this location is becoming more severe or not. Then, we will present the trend in the waterlogging map. At last, we use a LBS (Location Based Service) server from Baidu[3]. After uploading all those locations, Apps can easily get those data through the network and show those points on the map.

To sum up, our approach is able to retrieve waterlogging locations and make a prediction of the development trend.

3   Evaluation

3.1   Evaluation Goal

In this section, we will demonstrate our web site system and Apps. Our system can track the urban waterlogging in more than ten big cities in China. We will show the waterlogging map with severity and trend.

3.2   Evaluation Step

In order to create a demo for urban waterlogging monitoring system, some of the most serious time when severe urban waterlogging occurred in several big cities in China were manually selected for monitoring. Our system is able to get public posts through the Sina Weibo APIs. Table 1 shows some statistics about our data sources. We retrieve the data from July to November in the day when urban waterlogging happened in several big cities in China.

Firstly, we analyze the tweets and extract entities according to the LOAD ontology. Then all the locations and description of waterlogging will be analyzed

Table 1. Statistics of dataset

	Data Duration	Jul-Nov, 2013
Tweets	Sina Weibo	5246576
Reports	Waterlogging Reports	89

---

[3] http://lbsyun.baidu.com/

Fig. 4. Monitored waterlogging locations

by fuzzy set. We will evaluate the severity of each location and add tags to it. At last we will use the positive and negative examples to train the data and make a prediction of the development trend.

After analyzing the tweets, all waterlogging locations will be updated in LBS server, which will be pushed to Apps and showed website.

3.3   Evaluation Result

As showed in Fig. 4, our system provides a real-time picture of nearby waterlogging locations. Red locations on a map represent severe waterlogging locations and warn the user, while the blue ones represent the normal waterlogging. There is also a picture of a severe waterlogging. Moreover, we also provide users with statistics of urban waterlogging locations, which can provide detailed information about the whole urban waterlogging locations in history. Our system also allows users to add new urban waterlogging points with the description by themselves besides analyzing the social data. The system currently supports anonymously submission of urban waterlogging location's pictures or text. After processing the data in real-time, the new waterlogging location will be marked on the map right here to remind the user there has waterlogging occurred. Apart from its ability to function as a waterlogging monitor, we integrate this system with Sina Weibo client and Wechat client. When submitting the waterlogging coordinates, users also update their status in their Wechat[4] or Sina Weibo.

Our system has collected ten big cites's historic tweets related to the urban waterlogging. In total, we have calculated 1043 possible waterlogging locations and all these locations are being monitored. As showed in Table 2, our system has high precision through historic data analysis with a comparison of real locations.

---

[4] http://www.wechat.com/

Table 2. Urban waterlogging cities's location reported and monitored by our method in total in 2013

Cities	News reported waterlogging	Monitored waterlogging	Recall	Precision
Hangzhou	64	59	92 %	87 %
Beijing	98	85	86 %	86 %
Shanghai	133	101	75 %	77 %
Shenzhen	56	49	87 %	79 %

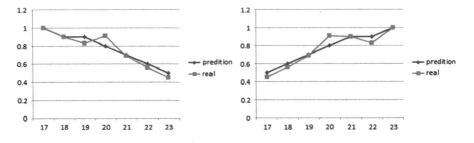

Fig. 5. Waterlogging trend prediction

We choose two test locations randomly in Beijing in July 23rd 2013 for predicting trends. In Fig. 5, the abscissa is time, and the ordinate is waterlogging severity in two locations. Our prediction results are quite close to the real situation.

4   Disussion and Future Work

Users share a great deal of content every day in tweets. People post large quantities of information associated with waterlogging when occurred. We fill this gap by creating a system which will provide users a rich set of statistics and visualizations of both real-time and historic urban waterlogging situation, allowing and warn our users to understand whether nearby roads are clear or not.

Retrieving knowledge from large datasets and dynamic networks is a challenge nowadays. Filtering high quality content, processing the data in real time, learning and predicting the new waterlogging locations, navigation in waterlogging areas are some examples of important tasks in which we could apply the technology we are building in this research project.

In future, we plan to enhance the system. We also plan to support twitter data to analyze city waterlogging in the whole world. As future research directions directly applicable to our platform, we propose studying the recommendation of contents and machine learning to identify waterlogging locations. New features like location search engine are also among our plans, which might also require some research. Moreover, we intend to open a start-up using our technology.

Acknowledgement. This work is funded by LY13F020005 of NSF of Zhejiang, NSFC61070156, YB2013120143 of Huawei and Fundamental Research Funds for the Central Universities.

References

1. Abdelhaq, H., Sengstock, C., Gertz, M.: EvenTweet: online localized event detection from twitter. Proc. VLDB Endowment **6**(12), 1326–1329 (2013)
2. Agichtein, E., Castillo, C., Donato, D., Gionis, A., Mishne, G.: Finding high-quality content in social media. In: WSDM '08: Proceedings of the 2008 International Conference on Web Search and Data Mining, pp. 183–194 (2008)
3. Albakour, M., Macdonald, C., Ounis, L., et al.: Identifying local events by using microblogs as social sensors. In: Proceedings of the 10th Conference on Open Research Areas in Information Retrieval, pp. 173–180. le centre de hautes etudes internationales d'informatique documentaire (2013)
4. Backstrom, L., Kleinberg, J., Kumar, R., Novak, J.: Spatial variation in search engine queries. In: Proceedings of the 17th International Conference on World Wide Web, pp. 357–366. ACM (2008)
5. Barbier, G., Liu, H.: Data mining in social media. In: Aggarwal, C.C. (ed.) Social Network Data Analytics, pp. 327–352. Springer, New York (2011)
6. Becker, H., Naaman, M., Gravano, L.: Learning similarity metrics for event identification in social media. In: WSDM '10: Proceedings of the third ACM international conference on Web search and data mining, pp. 291–300 (2010)
7. Becker, H., Naaman, M., Gravano, L.: Identifying content for planned events across social media sites. In: WSDM '12: Proceedings of the fifth ACM international conference on Web search and data mining, pp. 533–542 (2012)
8. Chen, L., Zhang, C., Wilson, C.: Tweeting under pressure: analyzing trending topics and evolving word choice on sina weibo. In: Proceedings of the First ACM Conference on Online Social Networks, pp. 89–100. ACM (2013)
9. Christakis, N.A., Fowler, J.H.: Social Network Sensors for Early Detection of Contagious Outbreaks (2010)
10. Crooks, A., Croitoru, A., Stefanidis, A., Radzikowski, J.: #Earthquake: Twitter as a distributed sensor system. Trans. GIS **17**(1), 124–147 (2013)
11. Jin, X., Wang, C., Luo, J., Yu, X., Han, J.: LikeMiner: a system for mining the power of 'like' in social media networks. In: KDD '11 Proceedings of the 17th ACM SIGKDD International Conference on Knowledge Discovery and Data Mining, pp. 753–756 (2011)
12. Kryvasheyeu, Y., Chen, H., Moro, E., van Hentenryck, P., Cebrian, M.: Performance of social network sensors during hurricane sandy (2014). arXiv:1402.2482
13. Li, C.: Ecohydrology and good urban design for urban storm water-logging in Beijing, China. Ecohydrology & Hydrobiology **12**(4), 287–300 (2012)
14. Lingad, J., Karimi, S., Yin, J.: Location extraction from disaster-related microblogs. In: Proceedings of the 22nd International Conference on World Wide Web Companion, pp. 1017–1020. International World Wide Web Conferences Steering Committee (2013)
15. Liu, Y., Alexandrova, T., Nakajima, T.: Using stranger as sensors: temporal and geo-sensitive question answering via social media. In: Proceedings of the 22nd International Conference on World Wide Web, pp. 803–814. International World Wide Web Conferences Steering Committee (2013)

16. Majid, A., Chen, I., Chen, G., Mirza, H.T., Hussain, I., Woodward, J.: A context-aware personalized travel recommendation system based on geotagged social media data mining. Int. J. Geogr. Inf. Sci. **27**(4), 662–684 (2013)

17. Sakaki, T., Okazaki, M., Matsuo, Y.: Earthquake shakes twitter users: real-time event detection by social sensors. In: Proceedings of the 19th International Conference on World Wide Web, pp. 851–860. ACM (2010)

18. Sakaki, T., Toriumi, F., Uchiyama, K., Matsuo, Y., Shinoda, K., Kazama, K., Kurihara, K., Noda, I.: The possibility of social media analysis for disaster management. In: Humanitarian Technology Conference (R10-HTC), 2013 IEEE Region 10, pp. 238–243. IEEE (2013)

19. Schade, S., Díaz, L., Ostermann, F., Spinsanti, L., Luraschi, G., Cox, S., Nuñez, M., De Longueville, B.: Citizen-based sensing of crisis events: sensor web enablement for volunteered geographic information. Appl. Geomatics **5**(1), 3–18 (2013)

20. Schröter, K., Kreibich, H., Merz, B.: Rapid flood loss estimation for large scale floods in germany. In: EGU General Assembly Conference Abstracts, vol. 15, p. 8798 (2013)

21. Tang, J., Wang, M., Hua, X.-S., Chua, T.-S.: Social media mining and search. Multimedia Tools Appl. **56**(1), 1–7 (2011)

22. Vieweg, S., Hughes, A.L., Starbird, K., Palen, L.: Microblogging during two natural hazards events: what twitter may contribute to situational awareness. In: Proceedings of the SIGCHI Conference on Human Factors in Computing Systems, pp. 1079–1088. ACM (2010)

23. Xu, G., Li, L.: Social Media Mining and Social Network Analysis: Emerging Research (2013)

24. Yadav, K., Chakraborty, D., Soubam, S., Prathapaneni, N., Nandakumar, V., Naik, V., Rajamani, N., Subramaniam, L.V., Mehta, S., De, P.: Human sensors: Case-study of open-ended community sensing in developing regions. In: 2013 IEEE International Conference on Pervasive Computing and Communications Workshops (PERCOM Workshops), pp. 389–392. IEEE (2013)

25. Yan, R.: Large-scale social media mining in facebook. In: WSM '11 Proceedings of the 3rd ACM SIGMM International Workshop on Social Media, pp. 21–22 (2011)

26. Yang, C.C., Yang, H., Tang, X., Jiang, L.: Identifying implicit relationships between social media users to support social commerce. In: ICEC '12 Proceedings of the 14th Annual International Conference on Electronic Commerce, pp. 41–47 (2012)

An Ontology-Based System for Generating Mathematical Test Papers

Jianfeng Du$^{(\boxtimes)}$, Xuzhi Zhou, Can Lin, Deqian Liu, and Jiayi Cheng

Guangdong University of Foreign Studies, Guangzhou 510006, China
jfdu@gdufs.edu.cn

Abstract. Automatic test paper generation is highly helpful in teaching and learning. In order to generate a test paper that covers as many knowledge points as possible, it is needed to discover knowledge points from exam questions. However, the problem of automatically finding knowledge points is seldom investigated in existing work. To fill this gap, this paper proposes an ontology-based method to discover knowledge points from mathematical exam questions. Accordingly, a system for automatically generating mathematical test papers is also proposed. It composes a test paper by solving a pseudo-Boolean optimization problem. Its practicality is demonstrated by a task of generating mathematical test papers from hundreds of postgraduate entrance exam questions.

1 Introduction

Test paper generation is crucial in teaching and learning. Teachers need to generate test papers to verify students' level of understanding in their courses, while students may also need to generate test papers for more targeted practice. Manually composing a test paper is time-consuming and even tedious, thus automatic approaches are highly desirable. Although the problem of auto-generating test papers has long been studied, most of existing work focuses on developing algorithms for generating test papers from exam questions, such as genetic algorithms [6] and memetic algorithms [7], under the assumption that full information about exam questions, such as which knowledge points an exam question covers, has been given. The degree of coverage of knowledge points is a crucial measure for the quality of a generated test paper. However, for almost all publicly available exam questions, the knowledge points that they cover are not provided. In order to generate test papers from these exam questions, it is needed to discover knowledge points beforehand.

We focus on the problem of automatically finding knowledge points in an exam question, which is seldom investigated in existing work. We do not treat this problem as a multi-label classification problem [8] and learn a model to classify exam questions, because the learning course requires a mass of training data while manually labeling exam questions is laborious and error prone. Instead, we resort to ontology-based approaches since an exam question often contains some technical terms that can be modeled as ontological terms. We target the domain

© Springer-Verlag Berlin Heidelberg 2014
D. Zhao et al. (Eds.): CSWS 2014, CCIS 480, pp. 237–244, 2014.
DOI: 10.1007/978-3-662-45495-4_21

of mathematical test papers in postgraduate entrance exams. We manually construct an OWL 2 RL [4] ontology for this domain. Based on this ontology, we develop a method for discovering knowledge points from exam questions. Moreover, by reducing the problem of generating a test paper to a pseudo-Boolean optimization problem [3], we develop a method for automatically generating test papers from mathematical exam questions. Finally, we develop a system for discovering knowledge points from exam questions and for generating mathematical test papers.

The remainder of this paper is organized as follows. Section 2 describes how we construct the math ontology. Section 3 describes how to discover knowledge points from exam questions based on the math ontology. Section 4 describes our method for automatically selecting exam questions to form a test paper. Before concluding, Sect. 5 describes our proposed system.

2   A Math Ontology for Postgraduate Entrance Exams

By looking up all technical terms in the mathematical exam syllabus of year 2014, we construct a math ontology expressed in OWL 2 RL, which is a tractable profile of OWL 2 [4]. We create 32 classes, including Character, Method, KnowledgePoint and 29 subclasses of KnowledgePoint, as well as two object properties hasCharacter and hasMethod, which link instances of KnowledgePoint to instances of Character and Method respectively. The 29 subclasses of KnowledgePoint correspond to all those knowledge points that need to be discovered from exam questions, but the class hierarchy among them only expresses a portion of hypernym-hyponym relations between math terms. For example, the term "square matrix" is a hyponym of the term "matrix", but this relation cannot be expressed by the class hierarchy as "square matrix" is not a knowledge point that needs to be discovered from exam questions, i.e., "square matrix" does not correspond to any class.

To fully express hypernym-hyponym relations, we create a third object property isA and introduce individuals to represent math terms. For the aforementioned example, since "matrix" is a knowledge point, we use individual iMatrix to represent the term "matrix" and introduce a class assertion Matrix(iMatrix) to declare that iMatrix is an instance of class Matrix. Meanwhile, we use individual iSquareMatrix to represent the term "square matrix" and introduce an object property assertion isA(iSquareMatrix, iMatrix) to declare that iSquareMatrix is a hyponym of iMatrix. We totally create 329 individuals for all math terms. Accordingly, we use the following axioms to complete the hypernym-hyponym relations and the links from knowledge points to characters and methods.

○   isA ○ isA ⊑ isA

(Meaning) isA is transitive, i.e., x is a hyponym of y if there is another individual z such that x is a hyponym of z and z is a hyponym of y.

○   ∃isA.KP ⊑ KP (for all subclasses KP of KnowledgePoint)

(Meaning) x belongs to knowledge point KP (i.e. is an instance of KP) if there is another individual y such that x is a hyponym of y and y belongs to KP.

○   isA ○ hasCharacter ⊑ hasCharacter

(Meaning) x has a character y if there is another individual z such that x is a hyponym of z and z has a character y.

∘   isA ∘ hasMethod ⊑ hasMethod

(Meaning) x has a method y if there is another individual z such that x is a hyponym of z and z has a method y.

We assume that the math expressions in an exam question has been converted to latex texts. We create a datatype property hasExpression to store regular expressions that match latex texts against individuals. Since all exam questions that we have collected are written in Chinese, we also use the annotation property rdfs:label to store Chinese labels of individuals. The values of the above two properties are used during the extraction of individuals from exam questions.

3 A Method for Discovering Knowledge Points

There are two phases for discovering knowledge points. In the first phase, we compute all assertions that can be entailed by the math ontology so that the next phase can directly work with the *complete ABox*, i.e. the assertional part of the ontology (ABox) with all entailed assertions added, regardless of axioms in the terminological part (TBox). In more details, we first translate all axioms in the TBox to function-free Horn rules by using the standard method for translating description logics [1] to first-order logic e.g. given in [2], since OWL 2 RL corresponds to the function-free Horn fragment of the description logic $\mathcal{SROIQ}$ [5]. Afterwards, we treat all assertions in the ABox as ground facts. By combining the function-free Horn rules translated from the TBox and the ground facts taken from the ABox, we obtain a function-free Horn program. Then the complete ABox is the unique least model of this function-free Horn program and can be computed by a standard bottom-up method. In the second phase, when given an exam question, we first extract individuals from the question, then compute the weights of knowledge points w.r.t. the extracted individuals, and finally pick knowledge points that have sufficiently large weights. More details about the second phase are given below.

Given an exam question, the individuals are extracted by matching the labels (stored as values of rdfs:label) against characters in the question and by matching the regular expressions (stored as values of hasExpression) against latex texts in the question. For matching a label against a character sequence, we use the well-known *edit distance*, i.e. the total cost for rewriting a character sequence to another one through three kinds of steps: inserting, deleting and substituting a character. The costs for inserting, deleting and substituting a character can be determined respectively by a small training set. We define that a label matches a character sequence if the minimum total cost for rewriting the character sequence to the label is not greater than the product of the length of the label and a user-specified threshold which can also be determined by training data.

Suppose L is the list of (possibly duplicate) individuals whose labels or regular expressions match some parts of the given exam question. We need to compute the weight of a knowledge point (i.e. a class) C w.r.t. L, denoted by

$w(C, L)$. It is defined as the sum of conditional probabilities $P(C|e)$ for all elements e in L, i.e., $w(C, L) = \sum_{e \in L} P(C|e)$, where $P(C|e)$ measures the strongness of the evidence that an exam question containing e belongs to C. Intuitively, $P(C|e)$ should be in direct proportion to the degree of connectivity of C and e in the RDF graph $\mathcal{G}$ constructed from the complete ABox $\mathcal{A}^*$ by neglecting all isA assertions, as the impact of the isA assertions has been absorbed in $\mathcal{A}^*$. Using this intuition we define $P(C|e)$ as $\frac{n(C,e)}{\sum_{C' \in \mathcal{C}} n(C',e)}$, where $\mathcal{C}$ is the set of knowledge points (i.e. classes) that need to be discovered from exam questions, and $n(C, e)$ is the number of different paths from e to C in $\mathcal{G}$. More precisely, $n(C, e)$ is defined as

$$I(C(e) \in \mathcal{A}^*) + \#\{e' \mid C(e') \in \mathcal{A}^*, \mathsf{hasCharacter}(e', e) \in \mathcal{A}^*\} +$$
$$\#\{e' \mid C(e') \in \mathcal{A}^*, \mathsf{hasMethod}(e', e) \in \mathcal{A}^*\},$$

where $I(E)$ returns 1 if the Boolean expression E is true or 0 otherwise, and $\#S$ denotes the cardinality of a set S. Let $w^*(L)$ denote the maximum $w(C, L)$ for all knowledge points $C \in \mathcal{C}$. We only pick those knowledge points C such that $w(C, L) \geq \theta \cdot w^*(L)$, where θ is a user-specified threshold. In practice, θ can be determined by a small training set through some experiments.

Example 1. Suppose the list L of individuals extracted from an exam question is $\{\mathsf{iQuadraticForm}, \mathsf{iRank}\}$, the complete ABox $\mathcal{A}^*$ is $\{\mathsf{QuadraticForm}(\mathsf{iQuadratic}$ $\mathsf{Form}), \mathsf{QuadraticForm}(\mathsf{iRank}), \mathsf{Matrix}(\mathsf{iMatrix}), \mathsf{Matrix}(\mathsf{iXMatrix}), \mathsf{Matrix}(\mathsf{iYMatrix}),$ $\mathsf{Matrix}(\mathsf{iRank}), \mathsf{hasCharacter}(\mathsf{iQuadraticForm}, \mathsf{iRank}), \mathsf{hasCharacter}(\mathsf{iMatrix}, \mathsf{iRank}),$ $\mathsf{hasCharacter}(\mathsf{iXMatrix}, \mathsf{iRank}), \mathsf{hasCharacter}(\mathsf{iYMatrix}, \mathsf{iRank})\}$, and the threshold θ is $\frac{3}{5}$. Then $n(\mathsf{QuadraticForm}, \mathsf{iQuadraticForm}) = 1$, $n(\mathsf{Matrix}, \mathsf{iQuadraticForm}) = 0$, $n(\mathsf{QuadraticForm}, \mathsf{iRank}) = 2$, and $n(\mathsf{Matrix}, \mathsf{iRank}) = 4$, thus we have $P(\mathsf{QuadraticForm}|\mathsf{iQuadraticForm}) = 1$, $P(\mathsf{Matrix}|\mathsf{iQuadraticForm}) = 0$, $P(\mathsf{QuadraticForm}|\mathsf{iRank}) = \frac{1}{3}$, and $P(\mathsf{Matrix}|\mathsf{iRank}) = \frac{2}{3}$. It follows that $w^*(L) = w(\mathsf{QuadraticForm}, L) = \frac{4}{3}$ and $w(\mathsf{Matrix}, L) = \frac{2}{3}$. Since $w(\mathsf{Matrix}, L) < \theta \cdot w^*(L)$, we only extract a single knowledge point $\mathsf{QuadraticForm}$ from the exam question.

4   A Method for Generating Mathematical Test Papers

Let $K(Q)$ denote the set of knowledge points extracted from an exam question Q. Besides $K(Q)$ we also use three other attributes of Q to determinate whether Q should be placed in a generated test paper. These attributes are respectively the type of Q (e.g., choice, fill-in-the-blank, essay or proof), the degree of difficulty and the degree of distinction. The type of Q, denoted by $p(Q)$, can be easily obtained. The other two degrees of Q can be computed by statistics. Suppose there are n students answering Q and scoring $s_1, \ldots, s_n$ on Q, respectively. Then the degree of difficulty of Q, denoted by $d(Q)$, can be defined as $s_Q - \frac{1}{n} \sum_{i=1}^{n} s_i$ where s_Q is the full score of Q, while the degree of distinction of Q, denoted by $q(Q)$, can be defined as the standard deviation of $s_1, \ldots, s_n$.

By considering that a generated test paper should cover as many knowledge points as possible, we reduce the problem of generating a test paper to a pseudo-Boolean optimization (PBO) problem [3]. A standard PBO problem is of the form "minimize $\sum_{j=1}^{n} c_j x_j$ subject to constraints $\sum_{j=1}^{n} a_{ij} x_j \geq b_i$", where $x_j \in \{0,1\}$, and a_{ij} and b_i are integers for all $i \in \{1, ..., m\}$. For this reduction, we assume that the degree of difficulty $d(Q)$ and the degree of distinction $q(Q)$ for every exam question Q have been scaled to integers. This assumption guarantees that all coefficients in the reduced PBO problem are integers.

Suppose there are n exam questions and m knowledge points. For the reduction, we introduce one 0-1 variable x_i for every exam question Q_i, where $1 \leq i \leq n$ and $x_i = 1$ in the solution of the reduced PBO problem means that Q_i is placed in the generated test paper. In addition, we introduce one 0-1 variable y_j for every knowledge point C_j, where $1 \leq j \leq m$ and $y_j = 1$ in the solution means that C_j is covered by the generated test paper.

The goal of the reduced PBO problem is

$$\text{minimize } -\sum_{j=1}^{m} y_j.$$

To complete the reduction, we append two sets of constraints to express the relations between the existence of exam questions and the coverage of knowledge points. The first set of constraints, shown below, expresses that a knowledge point is covered by the generated test paper if at least one exam question that covers this knowledge point is placed in the generated test paper.

$$y_j - x_i \geq 0 \text{ for all } i \in \{1, \ldots, n\} \text{ and all } C_j \in K(Q_i).$$

The second set of constraints, shown below, expresses that a knowledge point is not covered by the generated test paper if none of the exam questions covering this knowledge point is placed in the generated test paper.

$$\sum_{1 \leq i \leq n, C_j \in K(Q_i)} x_i - y_j \geq 0 \text{ for all } j \in \{1, \ldots, m\}.$$

We also add constraints according to user-specified parameters.

Users may specify how many exam questions of each type should be placed in the generated test paper. Let T denote the set of different types of exam questions and n_t denote the user-specified number of exam questions of type t. The following set of constraints expresses this specification.

$$\sum_{1 \leq i \leq n, p(Q_i)=t} x_i = n_t \text{ for all } t \in T.$$

In addition, users may specify the range of the average degree of difficulty (resp. distinction) of exam questions in the generated test paper. Suppose the range of the average degree of difficulty (resp. distinction) is set as $[d_l, d_u]$ (resp. $[q_l, q_u]$). The following four constraints express these settings.

$$\sum_{i=1}^{n} d(Q_i) x_i \geq d_l \sum_{t \in T} n_t. \qquad \sum_{i=1}^{n} d(Q_i) x_i \leq d_u \sum_{t \in T} n_t.$$
$$\sum_{i=1}^{n} q(Q_i) x_i \geq q_l \sum_{t \in T} n_t. \qquad \sum_{i=1}^{n} q(Q_i) x_i \leq q_u \sum_{t \in T} n_t.$$

5   The Demonstration System

We built a system[1] on top of RapidMiner[2], an open source platform for data
mining and analysis. The system provides two facilities, namely discovering
knowledge points from exam questions (using the method proposed in Sect. 3)
and generating mathematical test papers (using the method proposed in Sect. 4),
both of which were implemented as processes in RapidMiner.

The first process, shown in Fig. 1, is used to discover knowledge points from
exam questions. It mainly consists of six components. The first component "Read
Documents from Files" is a standard component in RapidMiner and is used to
load exam questions from text files. An individual exam question should be
stored in an individual file and will be loaded as a document. The second component "Documents to Data" is also a standard component that is used to
covert documents into data records. It will generate one record for one document with a new data field added to store the content of the document. The
third component "Load OWL Examples" is a self-made component. It is used
to extract the values of the data property hasExpression and the annotation
property rdfs:label for every individual in our constructed math ontology and to
extract all class names from the math ontology. The fourth component "Find
Knowledge Points" is also self-made, which implements the method proposed
in Sect. 3 for discovering knowledge points from exam questions. It will create
a new set of data records, where each record stores the identifier of an exam
question, a knowledge point that this exam question covers, etc. The fifth component "Select Attributes" is a standard component in RapidMiner and is used
to remove data fields that are not used later. The last component "Aggregate
by Concatenation" is a self-made component, used to generate a new set of data
records, where each record stores the identifier of an individual exam question
as well as the concatenation string for the set of knowledge points that the exam
question covers. The stored information will be used in the second process for
automatically generating mathematical test papers.

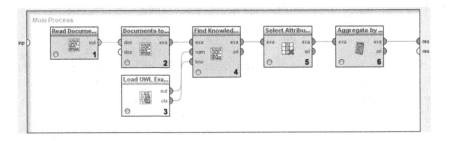

Fig. 1. The process for discovering knowledge points from exam questions

---

[1] More information about our constructed math ontology and our system can be found
at http://www.dataminingcenter.net/math/.

[2] More details about RapidMiner can be found at http://rapidminer.com/.

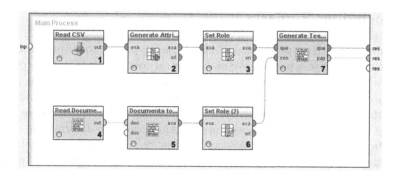

Fig. 2. The process for generating mathematical test papers

The second process, shown in Fig. 2, is used to automatically generate mathematical test papers. It mainly consists of seven components. The first component "Read CSV" is a standard component in RapidMiner and is used to load information about all exam questions from a text file. Each line in the given text file, converted into a data record by this component, should store the full information on an individual exam question, including the set of knowledge points that this exam question covers, the type, the degree of difficulty, the degree of distinction, the published paper and the sequential number of this exam question, where the values of different data fields are separated by some user-specified symbol. The second component "Generate Attributes" is also standard. It is used to compose an identifier for each exam question. This identifier should be the same as the identifier generated by the fourth component. The third component "Set Role" is also a standard component, used to set the role of the identifier field as "id". This setting is required for the "Generate Test Paper" component. The fourth component "Read Documents from Files", mentioned in the first process, is used to load exam questions from text files. It will generate an identifier for each loaded exam question. The fifth component "Documents to Data", also mentioned in the first process, is used to covert a document representing for an individual exam question to a data record. The sixth component "Set Role" is the same as the third component. The last component "Generate Test Paper" is the core of this process. It implements the method proposed in Sect. 4 for generating mathematical test papers. The implementation calls a highly optimized PBO solver MiniSAT$^+$ [3] to solve the reduced PBO problem. User-specified constraints can be provided as parameter values of this component.

To demonstrate how the two processes work, we collected 759 mathematical exam questions from postgraduate entrance exams in year 2003 to year 2013. We executed the two processes on a laptop with Intel Dual-Core 2.20 GHz CPU and 4 GB RAM, running Windows 7, where the maximum Java heap size was set to 512 MB. The first process finished in 108 seconds. The second process finished in variant time, from a few seconds to several hours, depending on different parameter values of the "Generate Test Paper" component. Note that the reduced PBO problem is an NP-complete problem, thus the execution time for solving it can sometimes be long. These experimental results show that our proposed methods are rather efficient and can be of practical use.

6   Concluding Remarks

In this paper we have proposed an ontology-based approach to automatically generate mathematical test papers. We made several contributions in this approach. Firstly, we constructed a math ontology expressed in OWL 2 RL based on a mathematical exam syllabus. Secondly, we developed a method for discovering knowledge points from exam questions based on the constructed math ontology. The knowledge points that an exam question covers are crucial in determining whether this question should be placed in a generated test paper. Finally, we developed an efficient method for generating a mathematical test paper. It reduces the problem of generating a mathematical test paper to a pseudo-Boolean optimization (PBO) problem and enables to call highly optimized PBO solvers to tackle the problem. We have built a system on top of RapidMiner to implement the proposed approach. Experimental results on this system demonstrated that the approach is practical for dealing with hundreds of postgraduate entrance exam questions. For future work, we plan to conduct user evaluation to verify the effectiveness of this system in helping undergraduates prepare the mathematical postgraduate entrance exam.

Acknowledgements. This work is partly supported by the NSFC grants (61375056 and 61005043), the Guangdong Natural Science Foundation (S2013010012928), the Undergraduate Innovative Experiment Projects in Guangdong University of Foreign Studies (1184613038 and 201411846043), and the Business Intelligence Key Team of Guangdong University of Foreign Studies (TD1202).

References

1. Baader, F., Calvanese, D., McGuinness, D.L., Nardi, D., Patel-Schneider, P.F. (eds.): The Description Logic Handbook: Theory, Implementation, and Applications. Cambridge University Press, Cambridge (2003)
2. Du, J., Qi, G., Pan, J., Shen, Y.: Approximating linear order inference in OWL 2 DL by horn compilation. In: Proceedings of 2012 IEEE/WIC/ACM International Conferences on Web Intelligence, pp. 97–104 (2012)
3. Eén, N., Sörensson, N.: Translating pseudo-boolean constraints into SAT. J. Satisfiability Boolean Model. Comput. **2**(1–4), 1–26 (2006)
4. Grau, B.C., Horrocks, I., Motik, B., Parsia, B., Patel-Schneider, P.F., Sattler, U.: OWL 2: the next step for OWL. J. Web Semant. **6**(4), 309–322 (2008)
5. Horrocks, I., Kutz, O., Sattler, U.: The even more irresistible $\mathcal{SROIQ}$. In: Proceedings of the 10th International Conference on Principles of Knowledge Representation and Reasoning, pp. 57–67 (2006)
6. Hu, J.-J., Sun, Y.-H., Xu, Q.-Z.: The genetic algorithm in the test paper generation. In: Gong, Z., Luo, X., Chen, J., Lei, J., Wang, F.L. (eds.) WISM 2011, Part I. LNCS, vol. 6987, pp. 109–113. Springer, Heidelberg (2011)
7. Nguyen, M., Hui, S., Fong, A.M.: Divide-and-conquer memetic algorithm for online multi-objective test paper generation. Memetic Comput. **4**(1), 33–47 (2012)
8. Tsoumakas, G., Katakis, I.: Multi-label classification: an overview. Int. J. Data Warehous. Min. **3**(3), 1–13 (2007)

The Twitter Observatory

Exploring Social and Semantic Relationships in Social Media

Qingpeng Zhang[1]([✉]), Bassem Makni[2], and James A. Hendler[2]

[1] Department of Systems Engineering and Engineering Management,
City University of Hong Kong, Kowloon, Hong Kong SAR, China
qingpeng.zhang@cityu.edu.hk
[2] Tetherless World Constellation, Rensselaer Polytechnic Institute,
Troy, NY 12180, USA
{maknib,hendler}@rpi.edu

Abstract. With the rapid growth of social media and the mature of Web Science research, there is a crucial need for a new generation of open analytics environment, which are able to observe and analyze what is happening in social media. To meet this need, we propose the Twitter Observatory, which is designed to collect data from Twitter in real-time, and enable the analytics of both social and semantic data. This paper introduces the development and implementation of a work-in-progress prototype of the proposed Twitter Observatory, as well as discussions of our ongoing research.

1 Introduction

The demand of the availability and interpretation of data, analytical methods, and computational infrastructure motivates the emerging of the Web Observatory [1]. As part of the Web Observatory effort, we have been working on the development of Social Media Observatories, which aim to collect and integrate the data from multiple social media platforms, and enable the analytics across platforms and domains [2]. A core part of the social media observatories is the Twitter Observatory, which generates a majority of the activities in the social media.

This paper introduces the development and implementation of the proposed Twitter Observatory based on Semantic Web technologies. The organization of this paper is as follows. In Sect. 2, we briefly introduce how we used RDF triples and a well-defined Twitter Ontology to model and represent Twitter data as RDF triples. The construction of a RDF graph database is also described. Then, in Sect. 3, a work-in-progress Twitter Observatory based on the RDF graph database is presented with descriptions of major functions. Last, we conclude the paper with discussions of ongoing and future research in Sect. 4.

Note: This work was done when the first author was affiliated with the Tetherless World Constellation at Rensselaer Polytechnic Institute, Troy, NY.

© Springer-Verlag Berlin Heidelberg 2014
D. Zhao et al. (Eds.): CSWS 2014, CCIS 480, pp. 245–250, 2014.
DOI: 10.1007/978-3-662-45495-4_22

2 The Graph Database Backend

In the Semantic Web paradigm, the world entities and relations are described using triples that follow the form of subject–predicate–object expressions, known as triples [3]. The triples are stored and managed via a graph database. In the graph, nodes represent entities and edges represent the relationship between represented entities. Therefore the notion of graph database naturally fits the social graph structure. In this project, we are using Semantic Web and graph databases to manage social media feeds. A core notion of the Semantic Web is ontologies, which allow the share of knowledge allowing the use of the stored triples [3].

The prototype system has collected millions of tweets for a number of selected topics and transferred them into high performance RDF graph database based on the Twitter ontology developed by our team[1] (Fig. 1). The ontology captures the common knowledge about social media networks formed by the use of Twitter. In addition, it also incorporates some key concepts of other social media, so that there is space to integrate data from other resources downstream. In summary, the ontology models the notions of blog, micro-blog, tweet, retweet, post, sentiment, etc. It is worth noting that we can always extend the Twitter Ontology to record other information like analysis results, new attributes, and other social media data.

Currently, the topics being collected by the system cover a variety of domains. The database is being fed dynamically with the ongoing real-time collection using Twitter Streaming API[2]. A SPARQL endpoint is constructed to allow the project members to query the data store, perform their analysis on the tweets, and write back the results of their analysis such as adding a triple indicating the sentiment of tweets, or attributing topological features.

3 Analytics and Learning on the Graph Database

Based on the Twitter Ontology introduced above, we have developed a Twitter "metadata" model that allows querying based on location, time, topic, user characterization, hash-tag characterization, sentiment, etc. A SPARQL endpoint has been developed for retrieving data with semantic-rich queries. The data, which supports the following functions, is based on the query sent to the endpoint. The implementation of the RDF graph database and the SPARQL endpoint is based on an open source RDF triple store – Virtuoso [4].

In the rest of this section, we introduce the existing functions of the proposed prototype system – the link of tweets to linked open data and the social network analysis.

3.1 Linking Tweets with Linked Open Data

We have been exploring existing and new tools to annotate social media data with linked data on the Web so as to exploit the rich information provided by the Linked

---

[1] The Tetherless World Constellation at Rensselaer Polytechnic Institute, http://tw.rpi.edu/.

[2] https://dev.twitter.com/docs/api/streaming

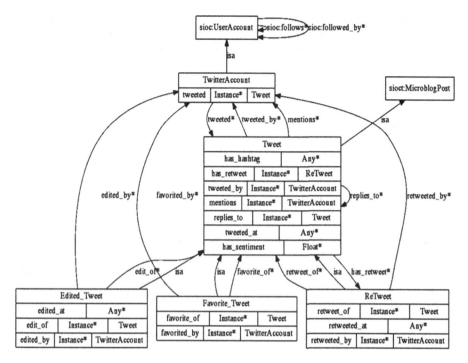

Fig. 1. The visualization of the Twitter ontology.

Open Data [5]. In particular, the current version of the Twitter Observatory employed the DBpedia Spotlight to identify and annotate the key concepts in tweets, and link it to the DBpedia entries [6].

To demonstrate the functionality of DBpedia Spotlight, we take the following tweet from the official twitter account of The Economist as an example – "A curious observation may lead to a treatment for multiple sclerosis http://econ.st/1kpkGyL"[3].

The DBpedia Spotlight annotates the following five DBpedia entries, highlighted in Fig. 2. In this example, the confidence value is 0.2, and the support is 20. As we can observe from the figure, the five annotations are "curiosity," "observation," "lead," "multiple sclerosis," and "hypertext transfer protocol." Among them, "hypertext transfer protocol" is a common annotation in tweets because of the nature of twitter sharing activities. Normally we exclude this type of common annotations. It's worth noting that the hyperlink itself will be identified and stored as well. For the rest four annotations, "multiple sclerosis" is the key topic of this tweet, and "curiosity" is the attitude towards the "observation" of "multiple sclerosis." There is an incorrect annotation of "lead," which points to the chemical element *Plumbum* (*Pb*). This is inevitable. In fact, one of our ongoing researches is the development of more reliable and efficient annotation algorithms to replace DBpedia Spotlight.

---

[3] https://twitter.com/TheEconomist/status/499873124834623488

The Twitter Observatory calls the Web service[4] of DBpedia Spotlight, and performs the annotation work automatically. Next, the DBpedia annotations will be stored into the database as new attributes (new RDF triples) of the corresponding tweet.

A curious observation may lead to a treatment for multiple sclerosis http://econ.st/1kpkGyL.
http://dbpedia.org/resource/Curiosity
http://dbpedia.org/resource/Observation
http://dbpedia.org/resource/Lead
http://dbpedia.org/resource/Multiple_sclerosis
http://dbpedia.org/resource/Hypertext_Transfer_Protocol

Fig. 2. DBpedia Spotlight annotation results of a typical tweet.

With the annotations, we are able to use the ontologies to explore the structured information in DBpedia. Figure 3 shows the screenshot of the DBpedia entry of "multiple sclerosis." There is much valuable structured information, for example, the DiseaseDB page of this disease (represented as "dbpedia-owl:diseasesdb ▪8412" in the page).

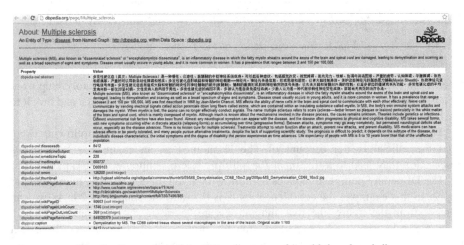

Fig. 3. Screenshot of the DBpedia entry of "multiple sclerosis."

Queries with these new attribute are enabled. We are exploring a novel scheme to integrate the social and semantic relationships for analytics. Please refer to Sect. 4 for a short description of our ongoing research. In addition, more efficient and reliable annotation tool is also part of our ongoing works.

---

[4] https://github.com/dbpedia-spotlight/dbpedia-spotlight/wiki/Web-service

3.2 Social Network Analysis

As the social groups and their activities in Twitter are essentially social networks by its nature, the social network analysis module is the key part of the Twitter Observatory, and the subsequent Social Media Observatories.

We developed a Web-based network analysis and visualization system to explore the relationships of people and semantics in the RDF graph database (Fig. 4). Users can visualize and analyze different types of sub-graphs based on the selections of topic, network definition, time range, sentiments, etc., which are supported by the Virtuoso data backend and SPARQL endpoint. The system performs a set of basic analyses (including degree distribution, centrality measures, clustering coefficient, shortest paths, etc.). These analyses enable the users to directly get the basic understanding of the organizational structure and dynamics of social groups in Twitter [7]. If the network is too huge to be analyzed by the system, or the users would like to perform a more specific analyses not provided by the system, the network can be exported to a number of popular network formats for offline use.

There is a set of most popular layout algorithms imbedded in the system for the visualization of networks. The visualization function is built with the Cytoscape Web toolkit [8].

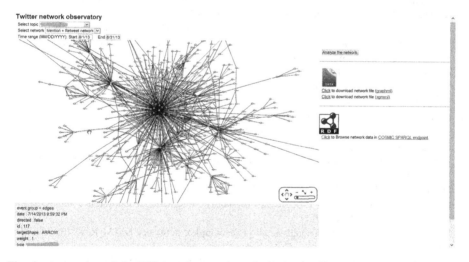

Fig. 4. A snapshot of the Web-based network analysis system (the topic has been hidden for privacy and confidential purposes).

4 Ongoing and Future Research

Based on the links to Linked Open Data, we are working on:

1. Developing more reasoning and analytics modules to inference on the semantic relations among the Twitter data. In particular, we are currently developing the function to visualize both social and semantic networks interactively, so that users

can perceive the information from both networks and the relationships between them.

2. Exploring the detailed interactions between the semantic-level network and the social-level network, with the goal to develop advanced algorithms to do mining on such heterogeneous networks. Models will be developed to describe the dynamics of the social groups with semantic information.

In addition, we are also working on developing a series of observatories that covers a broader range of social media and social groups (in particular, social media and social groups from multiple cultures). We plan to integrate the Twitter Observatory demoed in this paper with new observatories and develop a centralized Social Media Observatory, which will then become a crucial part in the worldwide Web Observatory.

Acknowledgments. This work is funded under the Rensselaer endowment to the Tetherless World Constellation.

References

1. Tiropanis, T., Hall, W., Shadbolt, N., De Roure, D., Contractor, N., Hendler, J.: The Web science observatory. IEEE Intell. Syst. **28**, 100–104 (2013)
2. Gloria, M.J.K., McGuinness, D.L., Luciano, J.S., Zhang, Q.: Exploration in Web science: instruments for Web observatories. In: Proceedings of the 22nd International Conference on World Wide Web Companion, International World Wide Web Conferences Steering Committee, pp. 1325–1328 (2013)
3. Allemang, D., Hendler, J.: Semantic Web for the Working Ontologist: Effective Modeling in RDFS and OWL. Elsevier, Waltham (2011)
4. Erling, O., Mikhailov, I.: RDF support in the Virtuoso DBMS. In: Pellegrini, T., Auer, S., Tochtermann, K., Schaffert, S. (eds.) Networked Knowledge - Networked Media. SCI, vol. 221, pp. 7–24. Springer, Heidelberg (2009)
5. Bizer, C., Heath, T., Berners-Lee, T.: Linked data-the story so far. Int. J. Semant. Web Inf. Syst. **5**, 1–22 (2009)
6. Mendes, P.N., Jakob, M., García-Silva, A., Bizer, C.: DBpedia spotlight: shedding light on the Web of documents. In: Proceedings of the 7th International Conference on Semantic Systems, pp. 1–8. ACM (2011)
7. Newman, M.E.: The structure and function of complex networks. SIAM Rev. **45**, 167–256 (2003)
8. Lopes, C.T., Franz, M., Kazi, F., Donaldson, S.L., Morris, Q., Bader, G.D.: Cytoscape Web: an interactive web-based network browser. Bioinformatics **26**, 2347–2348 (2010)

Author Index